Professor DIETER HELM is Professor of Economic Policy at the University of Oxford and Official Fellow in Economics at New College, Oxford. He specialises in the environment, notably in climate change, biodiversity, water, energy and agriculture. Previous books have included *Green and Prosperous Land: A Blueprint for Rescuing the British Countryside*, *Burn Out: The Endgame of Fossil Fuels*, *The Carbon Crunch* and *Natural Capital: Valuing the Planet*.

In December 2015, Dieter was reappointed as Independent Chair of the Natural Capital Committee. He is also an Honorary Vice President of the Berkshire, Buckinghamshire and Oxfordshire Wildlife Trust.

Praise for *Net Zero*

'Clearly written, on an important subject, by someone who knows their stuff' Mark Avery

'You should read it' Julian Glover, *Evening Standard*

'Dieter Helm is one of Britain's foremost experts on energy economics and he has written a terrific book on the next agenda item once the COVID emergency has passed ... A fine overview of our climate policy failures and the options for doing better' Colm McCarthy, *Sunday Independent*

Praise for Dieter Helm's *Green and Prosperous Land*

'One of the most important books of the decade' *Country Life*

'Refreshingly straightforward . . . A pure economic argument for conservation' *Sunday Times*

'This is an important analysis, argued with passion, intelligence and rigour. It is timely too, because – as Helm makes compellingly clear – of the urgency of the problem' *Financial Times*

'A trenchant manifesto for change' *Nature*

'Delivers handsomely on the promise of its title' *New Scientist*

'A brave and forthright attempt to begin a new conversation' *British Wildlife*

Net Zero

HOW WE STOP CAUSING CLIMATE CHANGE

Dieter Helm

WILLIAM
COLLINS

William Collins
An imprint of HarperCollins*Publishers*
1 London Bridge Street
London SE1 9GF

HarperCollins*Publishers*
1st Floor, Watermarque Building, Ringsend Road
Dublin 4, Ireland

WilliamCollinsBooks.com

First published in Great Britain by William Collins in 2020
This revised William Collins paperback edition published in 2021

2022 2024 2023 2021
2 4 6 8 10 9 7 5 3

Copyright © Dieter Helm 2020

Dieter Helm asserts the moral right to be identified as the author of this work
in accordance with the Copyright, Designs and Patents Act 1988

A catalogue record for this book is available from the British Library

ISBN 978-0-00-840449-9

All rights reserved. No part of this publication may be reproduced,
stored in a retrieval system, or transmitted, in any form or by any means,
electronic, mechanical, photocopying, recording or otherwise,
without the prior permission of the publishers.

This book is sold subject to the condition that it shall not, by way of trade
or otherwise, be lent, re-sold, hired out or otherwise circulated without the
publisher's prior consent in any form of binding or cover other than that in
which it is published and without a similar condition including
this condition being imposed on the subsequent purchaser.

Typeset in Lyon Text by Palimpsest Book Production
Limited, Falkirk, Stirlingshire
Printed and Bound in the UK using 100% Renewable Electricity
at CPI Group (UK) Ltd

This book is produced from independently certified FSC™ paper
to ensure responsible forest management.

For more information visit: www.harpercollins.co.uk/green

To Susie, Oliver, Laura, Amelie and Jake

CONTENTS

PREFACE

I thought I had finished writing about climate change a while ago, with my two books on the subject – *The Carbon Crunch: How We're Getting Climate Change Wrong – And How to Fix It* and *Burn Out: The Endgame for Fossil Fuels* – and 'The Cost of Energy Review' (the Helm Review) I undertook for the UK government in 2017.[1]

Back in 2012, in *The Carbon Crunch* I asked the question: why has so little been achieved? I wanted to puncture the complacency, and especially the peak oil fantasy – that we were going to run out of fossil fuels, and hence the oil price was heading north, making what looked like expensive renewables cheap by comparison. I followed this up in 2017 with *Burn Out*, pointing out that we have enough oil and gas to fry the planet many times over, and that the exit from fossil fuels will be messy for the great oil-producing countries, and messy for the renewables, as the trend price of oil and gas falls back. In both books, I stressed that a big part of the answer lay with technical progress and new technologies, and suggested that current renewables technologies would not be enough to solve the problem whatever their initial contributions, and they would not be 'in the money' any time soon.

And so it has come to pass. The predictions of the peak-oilers have turned out to be nonsense, the price of oil (and gas) has fallen back and, whatever their advocates claim, renewables are not yet subsidy-free once all the costs have been taken into account.

What I had not anticipated was that no serious progress would yet have been made on the fundamental problem, and that the concentration of carbon in the atmosphere would still just keep on going ever upwards, without so much as a blip, and, if anything, slightly accelerate. Even the Covid-19 coronavirus pandemic has failed to puncture the rise in the concentration of carbon in the atmosphere, even if it has made a temporary dent in emissions. When set against the enormity of the consequences of climate change, the only rational response is anger.

If this failure to achieve anything much in the last 30 years had been the consequence of not trying, it would be bad but at least understandable. But this is not the case: a huge amount of political capital and money has been spent in the name of mitigating climate change. Many people have been led to believe that current policies are working and that we are making good progress. They are not and we are not.

It is hard to stand on the sidelines and watch the results of badly designed policies and the misleading of people whose goodwill in the face of the wall of costs they will have to pay is essential to making progress.

What really bothers me, and makes me pick up the pen again to write on climate change, is the widely repeated claim made by the UK's Climate Change Committee (CCC) in the foreword to its recent report 'Net Zero: The UK's Contribution to Stopping Global Warming':

> By reducing emissions produced in the UK to zero, we also end our contribution to rising global temperatures.[2]

This is misleading in that the net zero target the CCC advocates is for territorial emissions *in the UK*, and unless every other country we trade with gets to net zero by 2050 too, or we stop importing carbon-intensive goods, it is simply not true. Neither contribution is likely to be met, COP26 notwithstanding.[3] The net zero carbon production target takes no account of the carbon we import, and which pervades much of our spending. For a deindustrialised economy with only 20 per cent manufacturing, this is particularly pertinent. To give a simple example, if British Steel had closed down (something which was under active consideration at the time the CCC published its report), UK territorial emissions would have gone down, but the subsequent increased imports of steel from, say, China would mean global warming would go up, and by more than it would have had that steel manufacturing remained here. According to the CCC's logic, why not close the rest of the British car industry, and INEOS's Grangemouth petrochemical plant too?

Buried later in the CCC report (page 106) is a chart showing the real villain of the piece, carbon consumption, stated as some 70 per cent higher than carbon production (albeit both badly measured and incomplete). If the UK wants to make no further unilateral contribution to global warming, which as I shall argue it should, then it is the altogether harder *net zero carbon consumption* that matters, and not just the easier bit of net zero carbon *production*.

It is also misleading because net zero does not actually mean 'reducing emissions to zero in the UK'. It is not *gross* zero but *net* zero. Net zero means that natural and industrial sequestration must be equal to or greater than the remaining carbon emissions in 2050, of which there are still likely to be quite a lot under even the most optimistic scenarios.

The public are being misled, and the CCC's formulation has the neat consequence that it makes the climate change problem look like it is all about production (and therefore all about what business does), and not about our personal spending. Having set up this illusion, the CCC effectively admits this later on, suggesting that we will have to eat less meat, change the land cover by planting lots of trees, and alter our lifestyles quite a lot. Here it is on stronger ground, although, as we shall see, not by any means strong enough.

It is not just misleading the public that angers me. It is also that a lot of time has been wasted. Thirty years on from the UN's drive to address climate change, we are still going backwards at an alarming rate. Up to the start of 2021, the concentration of carbon in the atmosphere carried on going up relentlessly at around 2 parts per million (ppm) every year. Not even Covid-19 lockdowns made a dent in the rising concentrations, even as they sharply reduced GDP and emissions.

The CCC compounded the misleading of the public when it published its Sixth Carbon Budget at the end of 2020.[4] It claimed that the costs of meeting the net zero carbon production target would be limited to 1 per cent GDP, or even less. This was the headline it chose to highlight in the media. Not mentioned was the assumption that this would result only if all the policy measures were implemented in a timely way. There is no government failure here, notwithstanding the fact that, to date, large-scale government failure has been the norm rather than the exception. The public now believes we are making good progress and it is not going to cost them much. No sacrifices here – just the Prime Minister's 'cakeism'.

Anger is not enough, and neither is despair at what has so far failed to happen. We can do much better. There needs to be a plan. This is my attempt to bring together my earlier arguments and analyses, to set out a better way of thinking through the carbon problem, and to lay out what a carbon policy would look like if we really wanted to limit global temperatures. It is what I think we should do.

This is about you and me, and about particular countries like the UK. The global UN-led process has so far failed. Kyoto and now Paris have not made any real difference, and indeed to the extent that political leaders who signed their countries up to Paris tell their voters and citizens that they are therefore taking action, their pledges can become fig leaves for business as usual. They (and us) can be seduced into thinking that it is 'job done'. Wishing the end entails willing the means.

I am not against more jaw-jaw at the jamborees that the annual Conference of the Parties (COP) have become. But I am against placing much faith in them producing results quickly enough to head off what is likely to happen, or even any meaningful results at all. What focusing on carbon consumption rather than just production does is put the spotlight on the polluters – you and me – and get the polluters to do something about their polluting ways. You and I can make a difference if we change our carbon-consuming habits, and this will make a difference if countries take up this obligation on a carbon-consumption basis. Better still, it turns out that if those countries exporting carbon-intensive products face a carbon border tax on a level playing field with home producers, this encourages them to introduce their own carbon price and keep the money rather than pay it to our government. I will explain how all this works to get much more effective global action than Kyoto, Paris and Glasgow.

We are not impotent, even in the face of something so large-scale and daunting as climate change. Not only can we make a difference to emissions – we can and should reduce our carbon consumption to net zero – some of these changes can also make our lives better. It is pretty obvious that improving air quality by lowering emissions in cities is good for us. It should be good too in Beijing, Delhi and Lagos. But it gets better once the natural carbon sequestration is brought into the frame. Climate change is not just about what we put into the atmosphere, but what nature takes out.

This is the link to my other and more recent writings on natural capital and the use of the land in *Natural Capital: Valuing the Planet* and *Green and Prosperous Land: A Blueprint for Rescuing the British Countryside*.[5] I have been working away at how to ensure that the UK government's overall objective of leaving the natural environment in a better state for the next generation might be achieved in the 25 Year Environment Plan and the supporting legislation.[6] What I have subsequently realised is that how we use the land is a very big part of how we crack climate change, and that it is a key part of net zero. Instead of focusing exclusively on carbon *emissions*, we also need to take seriously the other side of the equation: the natural *sequestration* by the land, by the trees, soils and peats, of those emissions. Nature, if protected and enhanced, does this for free all the time, and in the process produces all sorts of other benefits for us too.

In *Green and Prosperous Land*, I set out the key principles that should motivate action on the natural environment. These are: the *polluter-pays principle*; the provision of *public goods*; and the *net environmental gain* compensation principle. It is these same principles that should motivate decarbonisation. The polluters – you and me – should pay, and that needs a carbon price to incorporate the costs of our pollution, and this should be applied to imported carbon too (and hence trade). The public goods must be provided by the State, even if private companies do the work. These include the crucial low-carbon infrastructures, and research and development (R&D). None of these will be adequately provided by the market on its own. After all the mitigation options have been exhausted, the residual carbon emissions should be offset, and most of this should be natural rather than the industrial carbon capture and storage (CCS) option (although this too has a role).

The neat feature of this merging of the themes of *Green and Prosperous Land* and my earlier *Carbon Crunch* and *Burn Out* is that the environment gets treated holistically. Not doing so creates a very real danger that the silo approach to carbon policies, to the exclusion of everything else, could cause lots of collateral environmental damage. Think what would happen if the British Forestry Commission had in the past been let loose with the objective of maximising carbon take-up. The result would have been lots of spruce and even eucalyptus trees, damaging biodiversity, destroying peat bogs and acidifying water courses. It would have been a disaster on an altogether greater scale than the great green conifer blots on the uplands landscapes that the Forestry Commission created in its first 100 years. A proper plan for decarbonisation should be a plan for the environment *as a whole* – for water, biodiversity, health and well-being, all within a net zero carbon consumption context.

Significant bits of this plan are *no regrets*: we should do them anyway irrespective of carbon. This includes the R&D and the infrastructures: R&D has lots of unanticipated benefits, and there are lots of new technological advances that will be essential to cracking the carbon problem. I will detail some of these possibilities as the book progresses. Our infrastructures are in such poor shape that no one would relocate to the UK to access them. They have to be replaced anyway, regardless of net zero. The same is true of much of Europe, the US, India and Africa. There should be net environmental gain, not least because of the great mental and physical health benefits, which are now rigorously documented in the scientific literature, and much else besides.

Whether any democracy would vote for all the policies I suggest remains to be seen. The reason for hesitation is that the real message that a focus on carbon consumption brings home is that the root cause of climate change is one that is uncomfortable: our unsustainable carbon lifestyles. If you and I pay for that pollution, and preferably through a carbon price, then our standard of living will be impacted. It is going to cost a lot, and the challenge is to get on a sustainable consumption path, and hence pursue sustainable economic growth, not gross domestic product (GDP). We will have to rebase our lifestyles. It is going to hurt in the short term in order to have the long-term benefits. For if we do not get back onto a sustainable path, the corollary is that our economies will not be sustained. It will happen, it is just whether it happens *ex ante* by our deliberate choice, or *ex post* when nature bites back. That is the democratic choice. Failure to act does not abolish the consequences of not acting. They cannot be escaped.

Climate change can be cracked, but not if we carry on deluding ourselves, and our political leaders carry on colluding in maintaining carbon illusions. What is needed is to shine a light on what is going on, and to set out how to address the problem. Whether it will be enough, faced with the enormous challenge of China, India and Africa doubling their GDP every ten to fifteen years, and hence each being potentially as much as four times bigger by 2050, remains to be seen. It is our moral duty to try, and to do the no regrets stuff first, and it is our leaders' duty to tell the truth. That is what I try to do here.

LIST OF ABBREVIATIONS

AI, artificial intelligence
BSE, bovine spongiform encephalopathy
CAP, Common Agricultural Policy
CBA, cost-benefit analysis
CCC, Climate Change Committee
CCS, carbon capture and storage
CEGB, Central Electricity Generating Board
CfD, Contract for Difference
COP, Conference of the Parties
CRISPR, clustered regularly interspaced short palindromic repeats
DECC, Department of Energy and Climate Change
Defra, Department for Environment, Food and Rural Affairs
DDT, Dichlorodiphenyltrichloroethane
DETR, Department of the Environment, Transport and the Regions
DNO, distribution network operator
EFP, equivalent firm power
EMR, Electricity Market Reform
EU ETS, EU Emissions Trading System
GDP, gross domestic product
GMO, genetically modified organism
GW, gigawatt
HS2, a planned high-speed rail project for England
ICT, information and communications technology
IEA, International Energy Agency
IEM, Internal Energy Market
IP, intellectual property
IPCC, Intergovernmental Panel on Climate Change

LED, light-emitting diode
LNG, liquefied natural gas
mbd, million barrels per day
MPC, Monetary Policy Committee
NDC, nationally determined contribution
NFU, National Farmers' Union
OPEC, Organization of the Petroleum Exporting Countries
ppm, parts per million
PV, photovoltaics
R&D, research and development
RAB, regulated asset base
SUV, sport utility vehicle
TWh, terawatt-hour
UNFCCC, United Nations Framework Convention on Climate Change
USO, universal service obligation
VAT, value-added tax
WTO, World Trade Organization

INTRODUCTION

It is not going well. Thirty years ago, world leaders vowed to address the new great challenge – global warming. Thirty years later, the concentration of CO_2 in the atmosphere continues to grow unabated. Over three decades, it went up from around 355 ppm to well over 400 ppm, rising by around 2 ppm per annum. In 2018, it went up by 2.7 ppm. Not even a blip for the global financial crisis dents the relentless upward path. Only the Covid-19 lockdowns and associated temporary collapse of global GDP have checked the growth of emissions, but not the growth of carbon concentration.[1] Global temperatures have already increased by nearly 1°C since the Industrial Revolution.

The world burns ever more fossil fuels. The last 30 years have been the golden age of the fossil fuels. Far more fossil fuels have been burnt in the past 30 years than in the entire nineteenth century. We have benefited from the cheap energy to fuel our cars, heat and power our homes and our digital equipment, and produce all the many things we are now convinced we need, from clothes, houses and flat-screen TVs, to fast food wrapped in plastics. Fossil fuels have transformed our lives and are a key reason why we have such a high standard of living, and why China and other emerging economies have been able to take so many people out of poverty. Everyone else wants what the first carbon economies have got. If you think we are kicking the fossil fuel habit, think again.

If the objective set in 1990 was to reduce emissions and limit global warming, it has been an utter failure. We are not and have never been on a path to a decarbonised future. Sufficient carbon-based capacity is already installed in the world's energy systems to bust the 2°C target.[2] The numbers for China alone are staggering: it has now exceeded 1,000 gigawatts (GW) of coal-fired electricity generation capacity (for a rough comparison, the total capacity for all types of generation on the British system is around 85 GW). Furthermore, China has been building another 150 GW at home (equal to Europe's total coal capacity, and not far short of that of the US) and promoting and financing coal projects abroad.[3] Those misguided enough to see China as leading the way to a greener energy future just need to look at these facts.

It is not just the emissions but the other side of the carbon balance too. While pumping out ever more carbon, we have simultaneously been reducing the ability of the natural environment to absorb it. The capacity of nature to mop it up by natural sequestration through absorption of carbon by trees, soils and peat has been decimated over the last 30 years. Brazil is accelerating the destruction of the Amazon rainforest (this major carbon sink is now being destroyed at the rate of 1 hectare per minute); the Mekong rainforest in Southeast Asia is threatened by huge Chinese dams upriver; and the future of Africa's largest rainforest and thus carbon sink in the Congo looks grim. Summer fires rage in the peat bogs of Siberia – another immensely important carbon sink. Malaysia and Indonesia keep clearing their rainforests for palm oil. Intensive agriculture has released the carbon from the soil (and decimated its biodiversity), and depleted the peat bogs – degrading these great carbon sinks. Oceans are being acidified, thus decreasing their ability to absorb carbon.

Why is it going so badly? Weren't the Kyoto Protocol and the Paris Agreement supposed to deal with this? Why has it all been such an utter failure?

The usual answers are all about the failures of world leaders and corporations. It has been all Trump's fault. It is down to the Chinese Communist Party's addiction to coal. It is those evil oil and gas companies and the coal mining corporations. With one more big heave, encouraged by global demonstrations, increasingly demanding targets will be pledged by the main polluters and the global fossil fuel industry will be pushed towards oblivion as investors flee from their stranded assets. We can sail off into a new post-Glasgow summit prosperity of a low-carbon transition, towards a net zero world by 2050. If we all declare a 'climate emergency' and sign up to net zero, all will be well, or at least a lot less bad.

To see why this is not going to crack climate change, the key thing to realise is that this is looking down the wrong end of the telescope. It looks at net zero emissions and production on a territorial basis and sees the solution to climate change as switching that production from high to low carbon in specific territories that adopt a net zero target. Get the producers to change their ways, and all will be well. Just get politicians to force them to do so.

While of course production has to decarbonise, there is much more to the climate change problem – and net zero – than this simplistic approach indicates. The right place to start is with *consumption* and us the *consumers*, and only then should we look to the producers. We have to remember that all this stuff that is made with fossil fuels, and thus causes carbon emissions, is made for us. When we temporarily stopped consuming so much during the Covid-19 lockdowns, such as shopping, eating out and travel, down went emissions.

In order to get a handle on the consumption side and our role in causing climate change, we need to take a close look at our own carbon consumption and understand not only how much climate change each of us is causing, but also what we can do about it. By understanding our personal carbon consumption, we can then work out how we can minimise our own impact, and what that means for our local community and therefore the collective responsibility for climate change. Net zero for us, for local government, and for a country like the UK as a whole requires net zero carbon consumption – and net zero carbon consumption requires attacking both the production *and* consumption sides.

Some areas of carbon consumption are pretty obvious, and figures have already been calculated for us. Take our cars, or our travel by planes. It is not hard to work out the staggering emissions from an aircraft flying, for example, from London to New York. This might amount to a little less than a tonne per passenger, compared with the average per capita emissions of carbon production in the UK of around 10 tonnes per annum (average carbon consumption is likely to be much higher). Similarly, the emissions from different types of cars are documented. There are apps you can consult. To end your contribution to climate change you would need to stop flying (at least until there are very low-emissions aircraft, sadly a long way off), and switch to electric cars and public transport, although even here some carbon emissions will remain. The electric car is very much more carbon-intensive to build than a conventional petrol or diesel one; the electricity is not – and probably never will be – completely carbon-free, and nor will the trains, buses and other forms of public transport. To be truly carbon-neutral, you will need to work out how to offset what you cannot mitigate. That will be expensive for you.

Yet this is just the tip of the carbon consumption iceberg. To get a much more complete picture of your carbon consumption, and hence what consumption for all of us will need to look like in a truly net zero world, let me suggest an exercise for you. Try writing down (perhaps with the help of one of the websites or apps) a carbon diary, your very own carbon diary. The idea is simple, even if the detail is complicated. This will give you your best guess at the carbon embedded in everything you do in your typical day, from the moment you wake up to getting back into bed. Think how your day starts. Go to the toilet, and the toilet paper is energy-intensive to produce. Check your emails and download a newspaper, and the energy-intensive servers kick in. Breakfast comes in packaging and contains lots of fertiliser-based foods. Pesticides are probably involved too. Then there is the milk and the cows and the methane, and that palm oil mentioned above. Fry some bacon and eggs or eat a croissant, and think just how much carbon might be involved. You will be wearing carbon-intensive footwear and clothing, much of it washed in a machine using carbon-intensive detergents and energy. All of this before you travel to work, go for lunch, and have a night out on the town. Your lifestyle – and everyone else's – is riddled with carbon. Put another way, our standard of living is utterly dependent on carbon, and it is this lifestyle as reflected in your diary which the developing world aspires to, and the extra 2 to 3 billion people coming along this century will want too.

The point of this is not to make you feel guilty, although you may well end up being very guilty. It is rather to reveal what the world will have to look like if we are to decarbonise. It is what would have to change on the consumption side, and it is way beyond replacing coal and gas power stations with wind turbines, solar panels and nuclear power stations, important as these steps are.

Try to imagine what your carbon diary would look like in 2050, supposing for a moment that net zero is achieved. Almost all of the carbon listed in your current diary would have to go, and anything left would have to be offset by measures to sequestrate it. By comparing the two, it is possible to recognise just how radical net zero is going to be. This is no walk in the park.

What your carbon diary will also tell you is that a lot of your consumption comes from overseas. It is not just home carbon emissions that matter. These carbon imports are the result of energy- and carbon-intensive production abroad. All that stuff 'Made in China' is for you – your mobile phone, laptop, flat-screen TV, household appliances, clothes and trainers. This stuff is all part of your carbon footprint. Other stuff, like palm oil and beef, might come from countries that are clearing their rain-forests to produce it, and hence the loss of the carbon sinks, the carbon leaking from the soils and the loss of biodiversity are all for you too. Many of the products in the supermarket contain palm oil, often under different names. The data servers are usually overseas.

Our kitchens and bathrooms are awash with chemicals; our clothes made from synthetic fibres; our food comes from fertiliser- and other chemically assisted agriculture; our houses are made with steel and cement; and an increasing number of gardens are covered with artificial turf. The list goes on. Even if you buy 'renewable' electricity, it has lots of embedded carbon in the turbines and the solar panels, and needs fossil fuels to back it up. Invest in an electric car, with an electric battery, and your carbon footprint will still remain a big one. Think of the cobalt in the battery, the materials in the car frame, bodywork and interior, and of course the tyres and brakes. An electric car might take twice as much carbon to produce than a conventional one.

My point is that this highlights just how unsustainable our lifestyles have become.[4] We are becoming addicted to a way of living in which flying is regarded as essential by many, and an aspiration for most. Even Greta Thunberg, in her noble efforts to get world leaders to take climate change seriously, had a support team flying backwards and forwards to the 2019 UN summit on climate change as she sailed in a 'zero carbon' yacht. They had, it was claimed, no option. Very few of us could afford the costs of Greta's yacht. She demonstrated that it could be done without many emissions, but not in a way the rest of us could emulate. Others could simply have chosen not to travel. The actor Emma Thompson flew 5,400 miles (emitting around 1.6 tonnes of carbon) from and back to California to be in the Extinction Rebellion protest boat in London's Oxford Circus, and Prince Harry used a private jet for his holidays, having preached the environmental message on climate change a few weeks earlier. What is wrong with fibre and video links? Wouldn't that have been a much better example to all of us? This is something that the Covid-19 pandemic has taught us – it can be done without all the travel.

Now you see where consumption comes in. Our demand leads

to production and supply. Without us consuming all these carbon-intensive goods and services, they would have no market. And once the focus shifts to consumption, a wider truth is revealed.

Climate change is pretty simple as a conceptual problem: it is about a small number of gases, and it does not matter where they are emitted. It is also about the ability of the environment to naturally sequestrate them – the 'net' bit in net zero is the carbon that we will keep emitting after we have done all the mitigation we can and hence must put back into the oceans, the trees, the plants and soils and into the ground. What we need is to stop the emissions, *wherever* they are occurring, and to encourage natural sequestration to mop up as much as it can.

Now the radical implication. It is not enough to clean up our own backyard. This does not stop us contributing to global warming. It is a fantasy, propagated by politicians, the CCC and some activists, that if we could only get to net zero for our own territorial emissions – for our carbon *production* – that would mean that we would have crossed the Rubicon and no longer be causing any further global warming. It is an extremely dangerous delusion.

Worse still, net zero only at home could actually increase global warming. This is arguably what has happened under Kyoto, which resulted in the Europeans obsessing about their own carbon production and theirs alone. Think how you would quickly get carbon emissions down in the UK or indeed the whole of Europe. The best place to start is closing down energy-intensive businesses that produce the most carbon. Close down the remaining steel, aluminium, petrochemicals and fertiliser factories. Stop making cement. Do all this and territorial emissions will fall. Drive up energy prices to make sure these businesses face a serious competitive disadvantage.

That is what has been going on since 1990. There are virtually no large energy-intensive investments in Europe at all now. For the existing factories, shifting car production abroad, moving petrochemical businesses to the US and closing down the remaining bits of the steel industry will all help to push territorial emissions down further. Making sure that the servers for the digital world are all overseas helps too.

As long as we keep consuming all this stuff, global emissions will almost certainly go up as a result of importing from abroad rather than producing at home. It makes climate change worse. Think of all that coal burnt in China; think of the heavily polluting shipping to get this stuff to our home ports.

Once carbon consumption is taken seriously, all sorts of other unsettling results follow. Take the great carbon villain of the piece, the US and especially Trump. The simple mantra that the Europeans are the good guys and the US the bad guys is not quite as black-and-white as it seems. The US has a thriving, energy-intensive manufacturing base. As cheap shale gas has come on-stream, it is not only switching from coal to gas, but it is also reshoring lots of energy-intensive businesses from China and elsewhere, and attracting European investments too. It is swapping energy-intensive imports, based largely on coal, for domestic production based on gas. It is sobering to reflect that the US could even have a better record than Europe on carbon consumption, and without any of the European policies and their costs. Higher European energy costs to pay for the renewables may actually have contributed to the European companies switching their investments to the US. Electing Joe Biden has not changed these fundamentals very much.

Looked at this way, the European and UK approaches, and all the excitement about net zero carbon production targets, are a lot less convincing as a solution to global warming. A genuine net zero policy on a unilateral basis makes sense only if it is net zero *carbon consumption*, and hence applies the carbon policies and prices to imports as well as to home production. But this means higher bills, and a correspondingly lower standard of living. All that imported stuff in your carbon diary would be more expensive. This is not going to be a painless, low-cost pathway.

Looking through the carbon consumption lens helps to explain why so little has been achieved since 1990 (and why the Covid-19 lockdowns had such a big impact). Emissions know no boundaries, and the problem can be cracked only if everyone does their bit. The search for a global treaty is a noble effort to solve the classic collective action problem, but so far it has produced little by way of actual reductions in emissions that would not have happened anyway.

Only the Europeans have been in the serious business of Kyoto and the efforts of the last 30 wasted years. They tried to set an example by starting to decarbonise their domestic production. By ignoring carbon consumption, the efforts on global warming have been largely in vain. If this poor outcome had been achieved at low cost, if it had created new global European renewables giants, and if it had avoided causing collateral economic damage, this might not matter too much.

Europe failed on all three counts. All the world's great renewables companies are outside Europe. Many of those wind turbines in the North Sea and those solar panels harvesting the sunshine in Germany have come from China. China had a massive export market created for it and paid for by the European consumers and producers, and as these extra European energy costs drove down European industrial competitiveness, China's export market to Europe for energy-intensive goods grew.

You might think that, notwithstanding this failure to date, it is only a matter of time before these shiny new renewables are going to come to the rescue and solve our climate change problems. Aren't renewables already grid-competitive against the fossil fuels, so won't everyone, including the Chinese and the Americans, be switching from fossil fuels anyway to remain cost-competitive against the Europeans?

Sadly, none of this is true. Current renewables, once *all* the costs have been properly taken into account, are not yet cost-competitive with the fossil fuels (and prices). Nor is nuclear. Worse, they are not going to be anytime soon. As fast as renewables costs are falling, so too are fossil fuel production costs, and although there is great and welcome progress in getting the costs of renewables down, they will require subsidies for some time to come. If you believed the lobbying, and believed that renewables are already grid-competitive, then you would be witnessing demonstrations outside parliaments demanding an end to renewables subsidies. Sadly, the cheap renewables cavalry is not about to turn up. Getting out of fossil fuels is going to be a lot harder and needs more subsidies, not fewer. There is no free lunch here.

The carbon consumption lens not only tells us why the efforts so far have been such an utter failure, but what needs to be done if we really want to tackle climate change, and have a stab at holding to 2°C warming. Consumption starts with us, at home. Instead of relying on the regular cycle of the annual UN COP, on Paris, Glasgow and their successors, we can start from the bottom up. We don't have to wait: we can unilaterally decide that we want to at least ensure that we no longer cause further climate change.

This means adopting a net zero carbon consumption target. Thinking back to your carbon diary, you will immediately realise that this is altogether more radical. It must include eliminating not only the carbon in the stuff made at home, but for imports too. That will hit your standard of living much harder, but it is what is necessary to get onto a sustainable economic growth path, underpinned by sustainable consumption. We are living beyond our climate means, and we have to reduce our carbon-based consumption, and only then can the contribution of ideas and new technologies put us on a stable growth path.

How do we do this? How do we design a sustainable economy? There are three key principles: the polluter pays; the provision of public goods; and net environmental gain. They are all good 101 economics and a sustainable economy needs to implement all three of them.

Of these, the polluter-pays principle is the most important. Carbon and other greenhouse gas emissions are all examples of pollution costs that should be incorporated into the economy, not ignored. Pollution has to be costed and these costs have to be integrated into every decision made by businesses and consumers. Making polluters pay is just good economics. To leave these costs out is inefficient.

Now the radical implication. If pollution costs are built into the economy, then the prices of all the carbon-intensive stuff in your carbon diary are going to go up, and some by quite a lot. That is why you will be worse off. You and I, the ultimate polluters, will have to pay the price of our carbon-intensive lifestyles. Less-carbon-intensive goods and services will now be relatively cheap and, over time, as we strive to get to a net zero personal carbon budget, we will do a lot of switching.

Reflecting this carbon cost in the prices means there must be a carbon price at the border for imports, as well as for domestically produced stuff. Put simply, there is no point in pricing pollution at home, only to ignore it for imports. That is how we got to the mess we are in, and encouraged offshoring the carbon pollution and closing down energy-intensive businesses in the UK and Europe. The only way to unilaterally tackle climate change, bottom up, is to have a carbon price that applies to all the various sorts of consumption. There has to be a carbon border tax.

The second principle is the provision of public goods. When it comes to climate change there needs to be low-carbon infrastructures that are fit for purpose. Climate change won't be addressed unless the transport, energy and fibre networks are designed for the sustainable low-carbon economy. This means full fibre and quickly to serve the emerging digital and electric economy. It means road and rail systems that embed smart charging and regional travel, and less aviation. It means decentralised smart energy networks and systems that incorporate storage and an active demand side.

Ironically, the good news is that the UK and many developed economies have infrastructures that are in very poor shape and need a major upgrade anyway. These new infrastructures will need to cope with the new technologies to make the transition to the low-carbon economy because the existing ones are not going to be able to do the job on their own.

Current wind turbines, current solar panels and current nuclear power stations are not enough. The first two are low-density, disaggregated generators, and the latter, in their current form, are expensive and hard to build. New technologies do not arise spontaneously. The good news is that R&D is comparatively cheap when compared with the rewards. A lot more of this vital public good is needed too.

The third principle, net environmental gain, is all about compensating for the damage caused by our consumption, and in particular dealing with the hard to avoid and hard to mitigate emissions. Net zero does not mean no emissions. It requires the residual carbon to be offset. Here is where carbon sequestration comes in. Some of this will be carbon capture and storage (CCS), but the real opportunity here is for natural carbon sequestration, the stuff nature does for us for free all the time. Net environmental gain entails net carbon gain.

These three principles, which underpin a sustainable economy and sustainable economic growth, have to be comprehensively applied. There needs to be a common carbon price across the whole economy and for imports; the infrastructures need to be economy-wide; and compensating for environmental damage should be required where and whenever damage is done.

Their application will have the biggest climate impacts on all the main polluting sectors – agriculture, transport, heating and electricity. There is good news here too. There are practical measures that can and should be taken now on the path to net zero carbon consumption. Agriculture should be sequestrating carbon, not emitting it. Carbon in the soils is good for productivity and good for biodiversity. Soils, which hold around four times the carbon of the atmosphere, should be husbanded, not depleted. Trees not only mop up carbon, but provide multiple other benefits, from biodiversity, water quality and flood prevention, to mental and physical health benefits. Peat bogs are fantastically valuable carbon stores, and restoration makes them even better.

A transport revolution is in the making. The internal combustion engine has to go. Electric cars are coming, and in due course they may be autonomous. Smart charging is also coming. Long-distance railways have the chance to displace aviation, at least regionally. Hydrogen via electrolysis offers the opportunity to do something about the gross pollution of shipping. All of these will improve our air quality too.

Transport and agriculture will be increasingly underpinned by electricity (even the hydrogen will be made from it). So too will heating. The future is electric, because the future is digital and everything digital is electric. The future is also electric because it is hard to envisage any other energy vector to get us to net zero. The good news is that, of all these sectors, electricity is probably the easiest to decarbonise. It will need new technologies, but right now just getting out of coal would be the single most important step towards decarbonising our world. Coal is really evil stuff, and the benefits of getting out of it globally go way beyond climate change. New solar technologies are coming, greatly increasing the efficiency with which we can capture the infinite supply of solar energy. There is no shortage of energy. Energy demand *per se* is not the problem. It is carbon-based electricity generation, carbon-based transport, carbon-based agriculture and carbon-based heating that are the problems.

Close in on carbon consumption and make us, the polluters, pay. Reorient our climate strategies and policies to the bottom up, and not the top-down world of Kyoto, Paris and Glasgow and the great jamborees that climate conferences have become. Jaw-jaw, yes, but don't expect it to crack the problem. Price pollution. Apply a carbon price to imports, both domestically and at the border. Invest in the twenty-first-century infrastructures to facilitate a low-carbon economy, and invest too in R&D to bring on the new technologies we are going to need.

Stop pretending and recognise the brutal facts about what has been going on for the last 30 years and why it has been such an abject failure. It is realism, not spin and fake optimism about progress and costs, that we need. Only then can we get going on what will work. Focus on the natural environment and especially on natural sequestration. Bring the trees, the soils, the peat and the marine world into the carbon play. Agriculture should be centre-stage, not an add-on held at bay by farmers lobbying for subsidies.

Do all these things and we can have a better natural environment, lead healthier lives, empower ourselves to do something, and make sure we really do leave better natural capital for the next generation. This is what sustainable economic growth is all about: it *can* be a green and prosperous land, but not if we go on as we are.

You may conclude that this is all unrealistic, that it is never going to happen, because we will not vote for politicians and political parties who can make it happen. You may be right, but then you would have to resign yourself to the conclusion that would follow – that climate change is not going to be addressed. What is set out here is what would have to happen to crack the problem. It calls for a different economic model, and a fundamental rethink of our consumption, of our environmental footprint. This book sets out how to do it – a plan that could and would actually work. Do this, and a country can be sure not only that it makes no further contribution to global warming, but also that it has embraced sustainable economic growth and has not harmed other aspects of the environment in the process. It would then be an exemplar for the world.

Sticking our heads in the sand, as we have been for the last 30 years, will not make climate change go away. Tackling it will seriously impact on our standard of living. We may think we have a choice, but we don't if we care about future generations.

PART ONE

30 Wasted Years

1
NO PROGRESS

Thirty years ago, Prime Minister Margaret Thatcher gave a speech to the UN which summarised the science as it was then understood and highlighted the need for global action to address the great 'new' climate change problem.[1] This speech is about as green as it gets for a politician. Although a great deal of subsequent research has refined that science and our understanding of the immense and subtle complexity of our climate, the main planks of the analysis were already in place back then, and they have remained remarkably robust ever since. The greenhouse effect is a bit of nineteenth-century science. The increase in greenhouse gases was already documented, and the step from increases in emissions to high temperatures was a simple and obvious one to make. By 1990, nobody could reasonably claim that they did not understand what climate change was and why it mattered. The failure to act has flown in the face of the knowledge of what that failure will mean for future generations.

The UN decided that immediate action needed to be taken on the reasonable grounds that there was enough evidence to act, and delay was only going to make things worse. There followed the groundbreaking UN Framework Convention on Climate Change (UNFCCC),[2] signed in 1992, committing the signatories to action and drawing on the reports of the already established Intergovernmental Panel on

Climate Change (IPCC). This in turn led to the Kyoto Protocol of 1997 and the Paris Agreement in 2015, with targets indexed back to the baseline of 1990.[3]

Apart from nuclear arms treaties, there has probably never been anything like the UNFCCC in human history. Hopes were high. The fact that the Soviet Union had imploded added to the sense that there was a new world order, capable of tackling this new and huge problem. The politicians declared a new dawn and promised to act. They have been promising ever since.

What was entirely absent from these groundbreaking global commitments was any sense of what was actually about to happen. The sobering truth is that, since 1990, the world has witnessed the golden age not of renewables but of fossil fuels, the environmental disaster that has accompanied (and been significantly caused by) China's rapid economic expansion on a scale never previously seen in economic history, and a relentless increase in the carbon concentrations in the atmosphere. It is very hard to detect any positive benefits which have followed the adoption of the UNFCCC, and, as we shall see, some of the subsequent UN-driven institutional architecture, such as Kyoto and Paris, may actually have made things worse.

What has happened since 1990?

The first and most important fact, which cannot be repeated often enough, is that global emissions are still going up (but for the Covid-19 temporary lockdowns) as they have been since 1990. The result is that the concentration of CO_2 in the atmosphere is now well past 400 ppm, having increased from around 355 ppm in 1990, and from about 275 ppm before the Industrial Revolution.[4] This is evident in the charts below, which should be deployed on the wall of every politician's office.

The key thing to note about the first chart (Figure 1) is that not

Figure 1 Global CO_2 atmospheric concentration: global monthly mean concentration of CO_2 (ppm)[5]

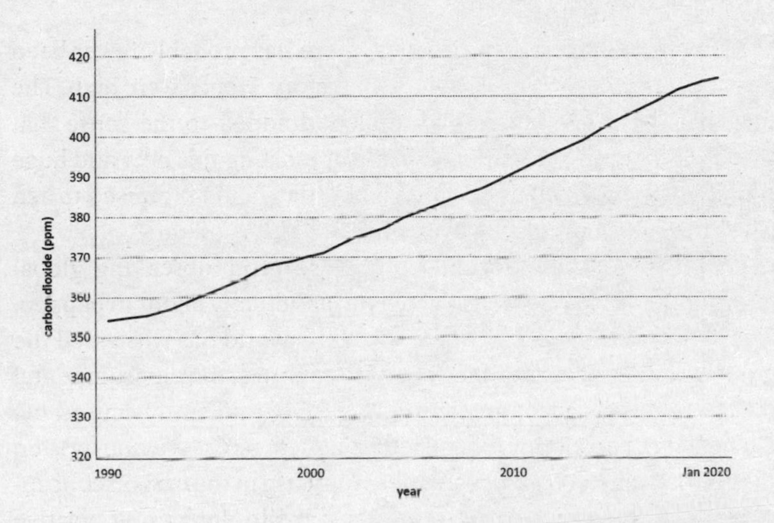

only are emissions still going up, but it is roughly a linear trend, and there are no blips in this trend. The great negative economic shock, the global financial crisis in 2007/08, does not even register. Nor do the effects of the Covid-19 pandemic.[6]

The 50 ppm increase in CO_2 since 1990 is around 60 per cent of the 75 ppm total increase since the Industrial Revolution (see Figure 2). The difference between the pre- and post-Industrial Revolution emissions is clear in the second chart. Just how 'special' the last 30 years have been can be seen by looking at the longer-term growth in CO_2 in the atmosphere, as illustrated in Figure 3.

There are three main reasons why emissions are still going up: burning ever more fossil fuels; the irrelevance of renewables so far; and the destruction of the natural environment's ability to sequestrate carbon.

Figure 2 Global CO_2 atmospheric concentration: global average long-term concentration of CO_2 (ppm)[7]

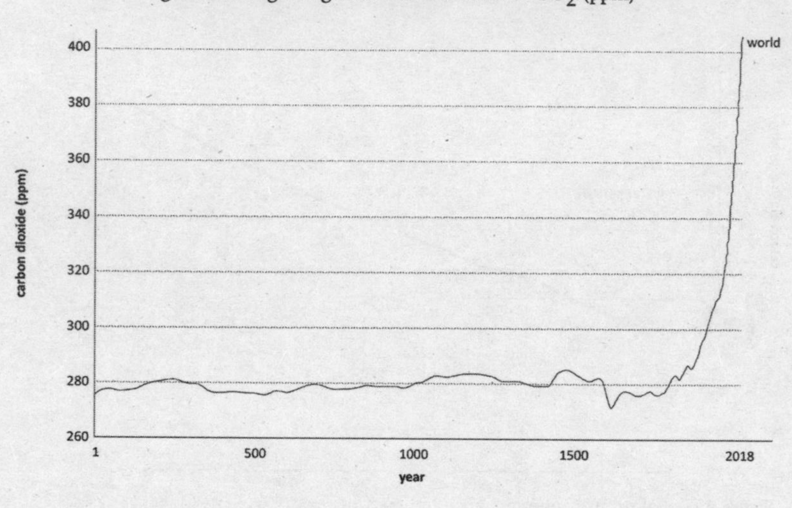

The golden age of the fossil fuels

The last 30 years have provided for a perfect fossil fuel combination: burgeoning demand, notably from China; the coming of gas and new technologies (notably shale); and the absence of any effective climate change policies. It could not have been much better from the perspective of the fossil fuel industries; from the climate perspective it could hardly have been worse.

The demand for each of the fossil fuels depends on its final market – and that is ultimately you and me. As illustrated in Figure 4, up until 1990 it was all about oil and coal. Gas was illegal to burn in power stations in the US and Europe until 1990, so it is very much a post-1990 fuel.

Oil is used for transport and petrochemicals. For all the talk of electric vehicles, almost all of the world's cars and trucks still use internal combustion engines, and, of these, diesel vehicles

Figure 3 CO_2 during ice ages and warm periods for the past 800,000 years[8]

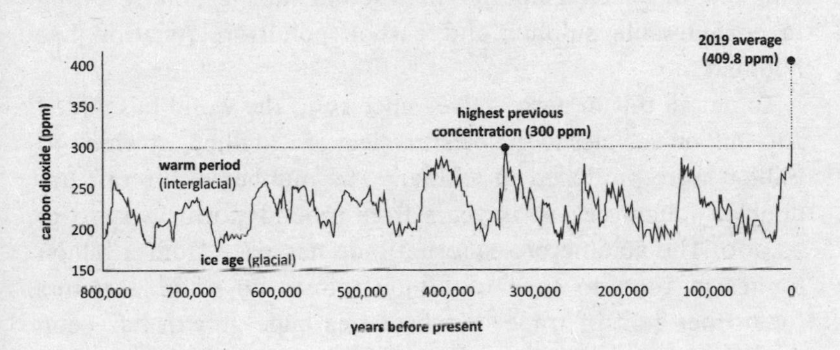

Figure 4 Global primary energy consumption,[9] measured in terawatt-hours (TWh) per year. Here 'other renewables' are renewable technologies not including solar, wind, hydropower, modern biofuels and traditional biomass.

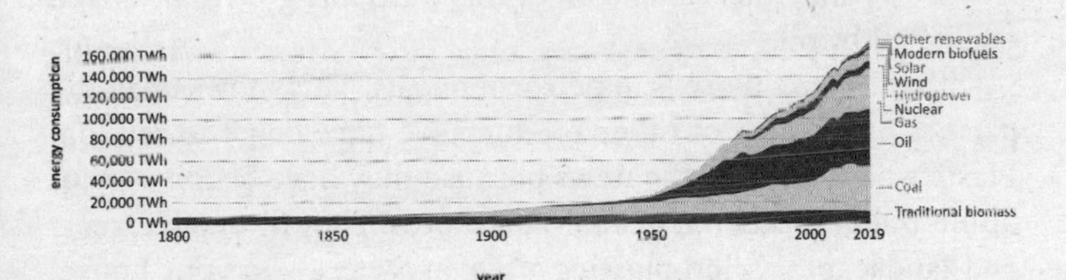

have seen the fastest growth. The growth of China has boosted the demand for cars, boosted world trade, and with it the shipping that has been using the dirtiest and cheapest oil, resulting in considerable sulphur and carbon pollution. Aviation has boomed too.

To put all this in perspective, since 1990 the world has added 100 million vehicles to the existing fleet of 1.2 billion, of which 30 million were produced in China.[10] The number of aircraft has doubled roughly every 15 years from around 5,000 in 1990 to 26,000. The volume of seaborne trade has risen from 4 billion tonnes in 1990 to over 10 billion tonnes now and accounts for 90 per cent of trade. Much of this huge growth has been container shipping, with an increase from 234 million tonnes in 1990 to 1,834 million tonnes in 2017.[11] This great transport growth has served to provide you and me with more car journeys, more deliveries to supermarkets and our homes, more flights for all those holidays, and all the goods that we buy that are packed into containers. We learned to appreciate just how much during the Covid-19 pandemic. Almost all of this transport growth has been facilitated by oil.

Oil's other main use is in petrochemicals. This shows up in the plastics and a host of other products we have come to rely on. Plastics have been in the news because they have become ubiquitous pollutants of our oceans and shores. They litter our streets and landscapes. Microplastics are everywhere – in your house, in your food, in your water – and they pollute every river, lake, seashore, and all the oceans, and can even be found in deep ocean trenches. You might naively believe you can go 'plastic-free', but that would simply reflect an ignorance of just how far our lives have become plastic-dependent. They are even inside you.

Petrochemical demand has grown strongly, especially in China. Petrochemicals are the main driver for the projected pre-pandemic increase in oil demand of around 10 million barrels per day (mbd) through to 2030, overtaking transport.[12] Demand will return to 100 mbd post-pandemic, and this oil-burn is obviously incompatible with the 2°C target. None of this is compatible with the 2°C target. It all fits into the continued growth of carbon concentrations in the atmosphere, extending the pattern since 1990.

Oil does not have the petrochemical market to itself. Gas has been making big inroads as its sheer abundance becomes apparent. It no longer needs to be treated as a premium fuel to be reserved for the higher-value petrochemicals it helps to produce. There is so much gas that its supply is best treated as effectively infinite, and it too is in its golden age. Gas also provides a good energy vector for heating and, increasingly, industrial processes. Your house is probably gas-heated, and you may cook with it too. Much of your electricity is now generated by gas-fired power stations. From 1990, gas became the fuel of choice for power generation in both the US and Europe, as the old reliance on coal gave way, and as nuclear stalled. In 1990 the typical electricity system in the US and Europe was around 80 per cent coal and 20 per cent nuclear. It is now much more likely to be 40 per cent gas and significantly less than 40 per cent coal. The closure of nuclear, through both ageing and anti-nuclear policies, has also made room for gas.

The switch to gas from coal for power generation reduces carbon emissions significantly – by around 20 and 40 per cent compared with oil and coal, respectively. The joker in the pack is methane and methane leakage, most of which comes from badly maintained pipelines and from leaking conventional and fracked wells. Methane leakage is troublesome because it is a very potent greenhouse gas – up to 80 times worse than CO_2 over 20 years,[13] although it does not stay in the atmosphere very long (and methane emissions from agriculture are more significant than those from natural gas).

Lots more coal since 1990

Coal is truly dreadful stuff. The story of coal since 1990 is remarkable, with its starring role in boosting carbon emissions. Chinese economic growth was (and still is) powered by coal, and this translates across to the growth of carbon emissions globally. In the 1990s growth in coal demand was rapid, but then began to level off at around 40 per cent of the world's primary energy. The chart below illustrates how coal consumption in China accelerated and ultimately exceeded that of the rest of the world.

Under any serious interpretation of the ambitions to address climate change, reducing coal should (and would) have been the primary achievement in the last 30 years. While there have been some notable successes, these are largely lost in the noise, with the exception of the coal-to-gas switch in the US. The

Figure 5 Coal demand in China and the rest of the world[14]

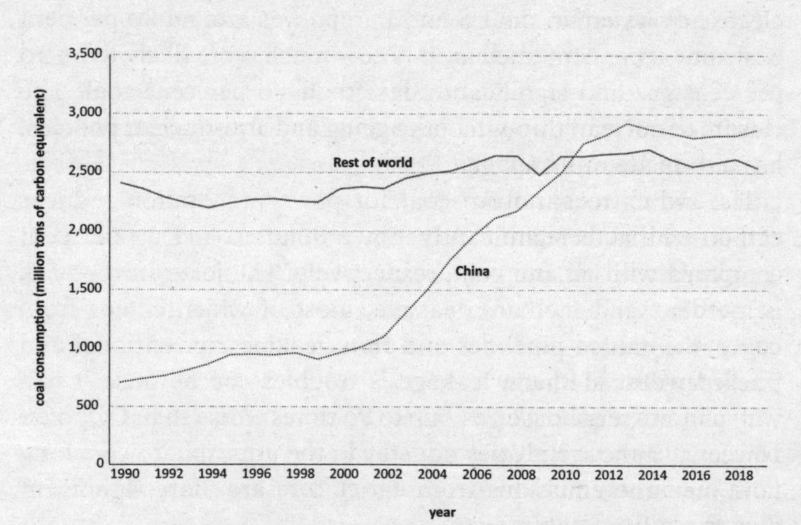

Figure 6 Coal production[15]

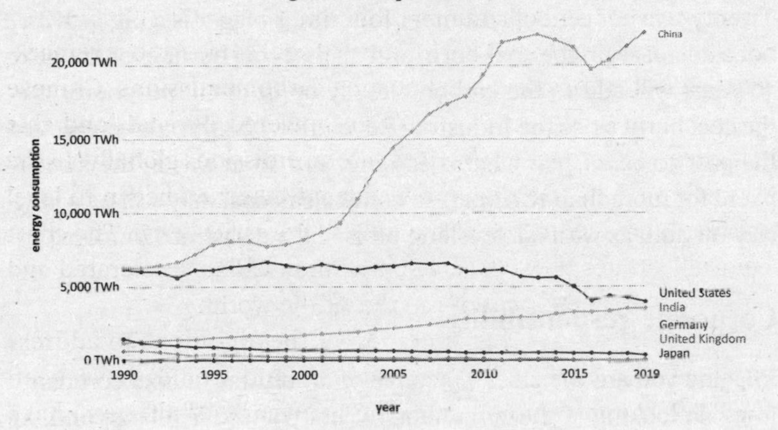

much-heralded UK squeeze on the coal industry, a rare example of successful carbon pricing, is just the flotsam of global coal-burn and therefore coal emissions. It takes only a few weeks for the world to replicate the entire coal generation capacity in the UK of a few years ago, and for most of the period from 1990 to now, the coal-burn held up in the UK anyway. The benefits of individual countries unilaterally closing coal come from the reductions of other pollutants (especially particulate matter PM 2.5 and its serious impacts on air quality), particularly close to cities, and to trees (notably sulphur and acid rain).

It is true that there are widely respected forecasts that coal's days are numbered, and from an environmental perspective this would be good news. But what is presented as a reduction in the *share* of coal in the energy mix is not the same as a reduction in coal *burnt*. The International Energy Agency (IEA) forecasts that, globally, coal will fall from around 38 per cent to 25 per cent by 2040 if all 'new policies' (those currently announced) are fully implemented.[16] Reflect, however, on how much higher global GDP will be by 2040 on current growth rates and as the world's economies recover from Covid-19.

As noted, China, India and Africa may be three to four times bigger. Twenty-five per cent of a number four times bigger is a big *increase*, not a decrease, in the coal-burn, and all that coal burnt over the next 30 years will add to the carbon already up in the atmosphere. It is the coal burnt since the Industrial Revolution *plus* the coal-burn over the past 30 years, *plus* what is coming, and all of this in the atmosphere for more than 100 years to come. Peak coal may still be a long way off and, as with oil, levelling off is still a disastrous outcome.[17]

Corporate responsibility

Imagine you are the chief executive of an oil, gas or coal company. Imagine looking at these numbers. What you see is a large and, in most cases, growing market over the next 20–30 years. You look at all the fossil fuel projects that cross your desk. You do the analysis. Perhaps you take a pessimistic view about medium-term future fossil fuel prices, given the sheer abundance of the resources available, which make a nonsense of all those peak oil and peak gas forecasts that were peddled by politicians and analysts up until the price collapsed in 2014 and then again in 2020.[18]

You will be able to balance some of the possible price falls with cost reductions as technology keeps marching on, and every bit as fast as for the competing renewables. Every day the understanding of the earth's crust deepens, and every day the drilling and extracting technologies improve. Soon you may have oil and gas platforms on the seabed available to you. The Arctic is being opened up by the Russians. What do you do? Do you say: 'I should stop exploration, stop new investments and wind down my business'? Or do you say: 'I should carry on and it is up to governments to impose a carbon price, regulate my company and take action'?

While some activists think you should do the former, their understanding of how global markets and boardrooms work is at best naive and partial. More than 90 per cent of these fossil fuels are produced by state-owned companies, and they are going to carry on. Why? Because their economies and their economic growth depend on them, and because they can. Think of Saudi Aramco and its centrality to the Saudi economy. Think of Russia's Rosneft.

And that is what they are doing, just like the tobacco companies, the manufacturers of sugary drinks, arms manufacturers, construction companies that use cement and steel, farmers who use fertilisers and pesticides, and so on. Notwithstanding the enthusiasm for environmental and social governance and disinvestment from fossil fuel producers, every company in the FT100 index is embedded in the fossil fuel economy. Those who say that this is what is wrong with the 'capitalist model' need to consider just what would happen if we jumped off now, rather than over a sensible transition, and why there is no effective carbon price. The statist model is, from a carbon perspective, much worse. It is the work of Saudi Aramco and Rosneft. Climate activists attack European and US politicians and company executives. They don't dare take on Vladimir Putin, Xi Jinping and Mohammad bin Salman. Gluing yourself to the HQ of Shell or BP is easy: doing it in Moscow, Beijing or Riyadh is much tougher.

The China factor

The growth of emissions since 1990 keeps coming back to China. The sheer scale of Chinese economic growth is unprecedented: nothing on this scale – or that of the environmental shock it has created – has ever happened before over such a short period of time. Only the Covid-19 outbreak has been able to make a serious dent, causing the economy to contract after decades of growth, and then only temporarily.

Think back to 1979. Most of the world's attention was on the Iranian revolution and the oil shock it produced. Nobody gave much thought to the impoverished Chinese communist country oppressed since 1949 by Chairman Mao. The Great Leap Forward (1958–62) and the Cultural Revolution (1966–76) left around 70 million Chinese dead from famine and the intellectual class decimated. China was a poor country, insignificant except for its nuclear weapons and its challenge to Russia for communist leadership.

Few noticed too the changeover at the top. Deng Xiaoping, like the rest of the ageing leadership, was a hangover from the Mao regime, lucky to have survived being purged. His great opening of the Chinese economy took time to gather pace, and the fear of a loss of Communist Party control hung over the experiments, as well as the people, who found their new hopes and opportunities crushed in Tiananmen Square in 1989.

China's GDP started to grow at around 10 per cent per annum, from its very low base, and it took until well into the 1990s for it to reach a size which would impact on global markets. When the Japanese economic miracle came to a shuddering halt in 1989, China was well placed to take up the slack, and gradually become the world's great exporter. The Chinese model was based, like that of Japan, on very high savings, which the State channelled into investments, and which also provided the money to lend to the US and others to pay for the Chinese goods they bought, and finance their consequent trade deficits. China would end up owning well over $1 trillion US Treasury bills, a sum Japan has also run up.[19]

The advantage China had over Japan was that its growth was initially built on cheap labour, an abundance of peasants migrating to the cities. It was an export-orientated expansion based on its east coast, and its products were precisely those which the West was getting out of producing: heavy, energy-intensive (and carbon-intensive) goods like steel. What made the impact of China all the greater was that it relied overwhelmingly on fossil fuels, especially coal.

In the late 1990s, China exported coal. Now it represents over half the world's coal trade (see Figure 6).[20] Coal-based heavy industry took over markets from Europeans and the US, and China's great environmental disaster started to unfold. The growth of China polluted the atmosphere and the nearby seas, impacted on the Mekong, and killed off much of the biodiversity in its other great rivers. The world's plastic problem is largely 'Made in China' too.

After 2000, China started to buy up agricultural land and raw materials globally, and after 2010 it launched its One Belt, One Road strategy (now known as the Belt and Road Initiative), financing infrastructure and power stations (especially coal) elsewhere.[21] Its shipping emitted lots of sulphur and carbon too. China literally changed the world's environment, and greatly boosted world trade and all its environmental costs. Put another way, had China continued to grow at its pre-1980 rates, there would be a significantly lower ppm carbon concentration in the atmosphere today, as well as more global and local biodiversity. Coal consumption would be significantly lower and trade would be significantly lower too. There would have been a much better chance of holding global warming at 2°.

Part of the reason for this massive effect is that emissions are concentrated in a small number of very large sectors, and these feature strongly in China's growth. China used a lot of cement in its industrialisation and the building of its great cities; it relies very heavily on fertiliser; and it has used a lot of steel domestically too. It has also exported all three of these, as well as petrochemicals.

The chart below is extraordinary in a chapter packed with the history of the great post-1990 growth. It shows the scale of the transformation. In 1990 Chinese GDP was around US$360 billion. By 2018 it was $13.6 trillion. That is *38 times bigger*. To get your head around these astonishing numbers, the UK's GDP in 1990 was just over $1 trillion, and it was just over $2.6 trillion before the pandemic.

With all this GDP growth comes pollution, and in China's case pollution on a truly planetary scale. Every other country in the world pales into insignificance in terms of added environmental pollution since 1990. The Europeans deindustrialised, and the US went sideways. From around 2005, the US had natural gas to substitute for coal, and hence it could both grow and limit its carbon emissions. On a carbon consumption basis, the net effect was relatively benign compared with what was going on elsewhere. China was off the scale in emissions.

Figure 7 The growth of China's GDP in trillions of US$ between 1960 and 2019[22]

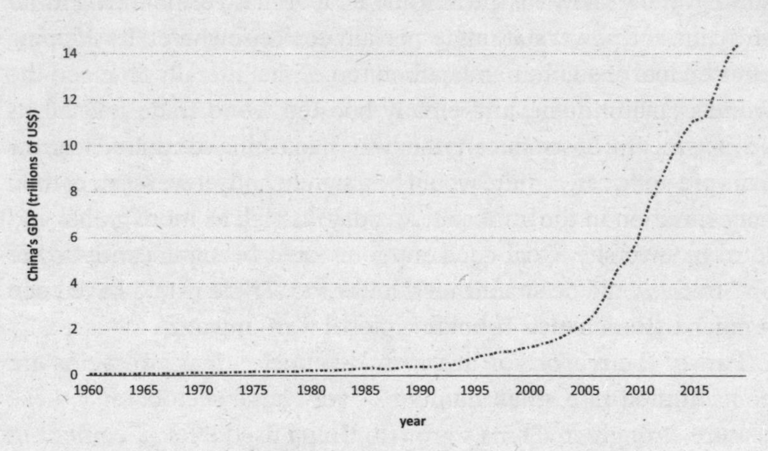

Table 1 CO_2 emissions by country, 2019.[23] Data is for CO_2 emissions from fuel combustion and cement production in tonnes in 2019.

Country	Total emissions	Per capita emissions
China	10.17 billion	7.1
US	5.28 billion	16.06
India	2.62 billion	1.91
Russia	1.68 billion	11.51
Japan	1.11 billion	8.72
Germany	0.70 billion	8.4
UK	0.37 billion	5.48

As the table above shows, China has almost twice the total emissions of the US, but less than half the per capita emissions. Unless it caps and reduces its emissions very quickly, the 2° target and the Paris Agreement mean very little. Nor is the Chinese economic expansion over. On the contrary, it was still growing in early 2020 at around 6 per cent according to official figures and, after an initial Covid-19 contraction, it was back on track by early 2021, and its emissions were again rising steeply. Just imagine if China eventually reached the US's per capita emissions.

Even if China does cap its emissions by, say 2030, India's projected economic growth and its reliance on coal provide further scary statistics. Coal currently accounts for over half of India's primary energy consumption, and its share might fall to a bit less by 2040.[24] But that might be for an economy four times its current size. In due course this could add another China to the global climate change problem. China could be just the first mega-example, followed by India and eventually Southeast Asia and Africa. Figure 8 shows the global top 20 highest emitters of CO_2 in 2018.

Figure 8 Top 20 CO2 emitters in 2018[25]

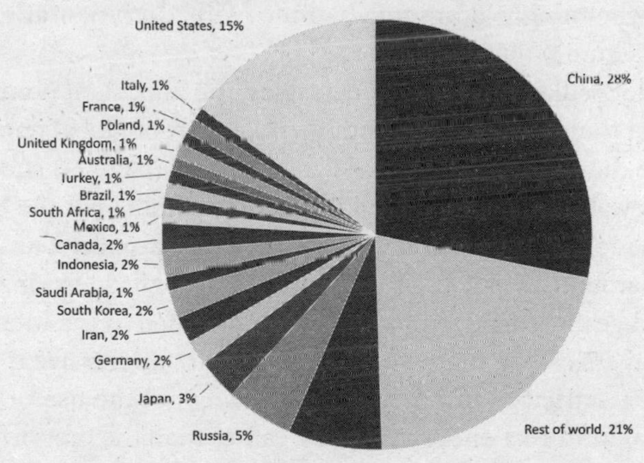

These are the fossil fuel facts. They describe the last 30 wasted years. The world's climate cannot withstand more Chinese coal-based economic growth and its replication in India and elsewhere. Against this juggernaut of emissions from the golden age of fossil fuels, renewables have, despite the hype, so far made little more than a minor scratch.

Renewables – noticeable by their absence

Back in 1990, the challenge was to replace fossil fuels with renewables, sources of energy which would not only be low-carbon but would keep coming at no extra marginal cost. Thirty years on and it still is.

At that time, there were four main renewable technologies, just as now: hydro, bioenergy, solar photovoltaics (PV) and wind turbines. Solar and wind are disaggregated, low-density and intermittent technologies. Storage, except through hydro, has not featured much in the last 30 years, and the demand side in electricity has been largely passive. Smart technologies and batteries have been small niche part-players. To date, hydro and bioenergy have overwhelmingly dominated, often with detrimental side effects, as Figure 9 illustrates.

For all the media hype about renewables, the important point to make is that they are still an incredibly small part of global energy supplies, to the point of being trivial, as Figure 4 (page 23) shows. Not all renewables are equal when it comes to carbon. For the UK, much is made up of biomass, and DRAX's wood pellet burning makes up around 15 per cent of renewables, with questionable anaerobic digestion using crops like maize included in the total.

The numbers can be made to look much more impressive if the focus is on electricity – in other words, leaving out the use of oil, gas and coal for direct energy supplies, rather than the proportion that goes into generating electricity. The much-hyped offshore

Figure 9 Modern global renewable energy consumption,[26] measured in TWh per year. Traditional biofuels refers to the consumption of fuelwood, forestry products, and animal and agricultural waste. Here 'other renewables' includes modern biofuels, geothermal and wave and tidal.

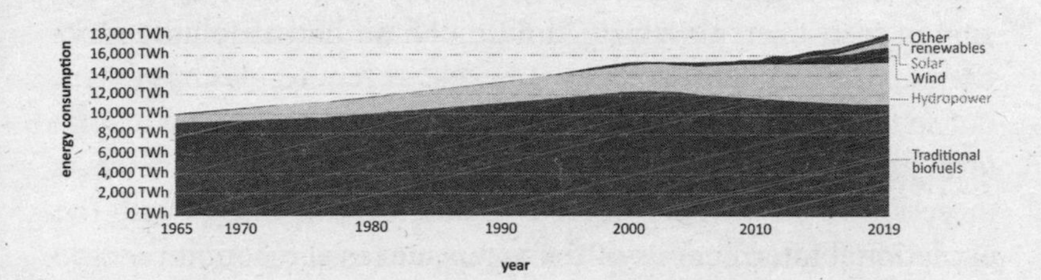

wind was just 0.3 per cent of global power supply in 2018.[27] Even this overstates the importance of wind (and solar too), since capacity and energy are not the same thing. Intermittency cuts down their effectiveness considerably (wind flows vary a lot). Solar is useless at night, and weak under much of northern Europe's cloudy skies. To date, wind and solar have needed fossil fuels to back them up, so for every unit of firm power added to electricity systems, the largest proportion of that unit continues to be fossil fuels in many cases.

It is true that substantial progress has been made on bringing down the costs of wind and solar, and increasing their technical capabilities. This has overwhelmingly been the result of heavy subsidies for mass production in China. There is a lot of scope in the next 30 years to push on with these cost reductions. But there are a couple of caveats. First, the cost reductions do not alter the fundamentals, and in particular the low density of energy that renewables produce; second, the fossil fuels have been showing remarkable technical progress too.[28] While the cost of solar panels has fallen, the cost reductions in shale and other non-conventional oil and gas, and in the conventionals, have been dramatic. Seismic technologies, robotics, artificial intelligence

(AI), and the digitalisation of the oil, gas and coal industries have all had dramatic impacts.

These considerations lead to a general conclusion: the cost of energy has fallen for all technologies. There has been little *relative* gain in costs for renewables. Almost all have had absolute reductions across all energy sources. The cost of renewables has fallen, but so too has the cost of oil. The oil price has maintained a long-term trend, gradually falling in real terms over the period since its inception as an industry in the late nineteenth century, with the exceptional interruptions of the 1970s and in the period running up to the peak in 2014 (and lots of short-term fluctuations). The renewables lobby is keen to argue that wind and solar are becoming grid-competitive. While they relied on a belief in peak oil this might well have turned out to be correct. But peak oil is nonsense: the renewables have not only to reduce their costs, but to do so *faster* than fossil fuels. The more successful renewables are, the lower the demand for oil, and hence the lower the cost of oil as production centres on the incredibly cheap sources in the Middle East. The marginal cost of Saudi crude oil is about US$5 or less, compared with a much higher world price, even after the price falls in 2014 and 2020. There is a long way to fall.

The destruction of natural capital

While the fossil fuels have boomed and contributed the lion's share of emissions since 1990, without much pressure from renewables, they are not the only cause of our climate problem. Climate change is a two-part act: the natural environment offsets carbon emissions with natural carbon sequestration through trees, vegetation, soils, peat and the marshes and oceans. The story here over the last 30 years is pretty bleak too – and, in a number of cases, because of the climate change policies we have naively been pursuing.

The fossil fuels on which we rely are themselves sequestered carbon, and for most of geological time this sequestration has been effectively permanent. This is ancient and stored solar radiation. Carbon is essential for photosynthesis and photosynthesis is what makes the planet work. It is also taken up by animals, as we can see in the seashells that litter our shores. Over time, these shells form the materials of the great limestone stores of carbon and what we use for cement.

Where these sequestration processes are allowed to play themselves out, a balance is established between oxygen and CO_2 in our atmosphere. The planet started out with little oxygen, and plants made our world habitable for animals, including ourselves. At times, oxygen was much more abundant, allowing super-sized animals to flourish.[29] In the great carboniferous age, oxygen was abundant precisely because, as the name implies, this was an age of great plant flourishing, which ultimately led to the formation of vast coal deposits.

The most obvious loss of this vital carbon-sequestering natural capital is the forests. They still comprise around 30 per cent of the global land cover, but we are cutting into this number at an alarming rate. Since 1990 roughly 1.5 million square kilometres have gone, including around 10 per cent of the Amazon rainforest. On average, an area of tree coverage the size of the UK was lost between 2014 and 2018 – 90 per cent of which was tropical forest.

Plants can both temporarily and permanently sequestrate carbon, and offset both the natural production of greenhouse gases and those emitted by our actions. Most of the current policies encouraging sequestration are temporary – and especially those implemented over the last 30 years – and many have been counterproductive. Planting a tree takes up carbon, and burning it releases it. Using biomass as 'renewable' energy closes the process. Over the life cycle of the plant, the best that can be achieved is carbon neutrality, but in practice producing biomass emits a lot of carbon. Why? Because not only does the planting and management require energy (and this tends to be tractors and fossil fuel-driven machinery), but it usually displaces something else, and that something else usually sequestrates carbon too. This is why we should be wary of treating 'bioenergy' as renewable.

Take the disaster that is the explosive growth of palm oil plantations to produce a 'renewable' fuel. Vast areas of natural rainforest are cleared. In Indonesia, so great are the fires that burn off the rainforests that many Southeast Asian areas suffer an annual smog. Attention has been paid belatedly to the impacts on biodiversity, but little attention has been paid to the net carbon accounting. In Indonesia, the underlying peat is set on fire too, adding to the environmental carnage.[30] The growth of sugar cane to produce ethanol in Brazil is another example of bioenergy which is less benign than it might seem. The sugar cane takes up land that had other uses, pushing out cattle farmers onto more marginal land. In Brazil's case this contributes to the destruction of the great Amazon rainforest. While Brazil's energy mix looks among the greenest in the world because of hydro, the reality is rather different. In the UK, anaerobic digesters use maize and rye grass to produce 'biofuels', with similarly dubious carbon-accounting justifications.

While the natural sequestration process of plants and soils should be soaking up carbon, agriculture (including forestry and other land use) contributes around 25 per cent of the world's greenhouse gas emissions.[31] Modern agriculture is a carbon disaster from start to finish, from the ploughing of the land releasing carbon stored in the soils, through to the final produce on the supermarket shelf, often wrapped in plastic. Ever since the discovery of the Haber-Bosch process to produce ammonia, fertilisers have been energy-intensive, and that energy has come overwhelmingly from fossil fuels. (China accounts for around 30 per cent of the world's total fertiliser use, with an application rate among the highest in the world.)

The statistics on what has happened to agriculture since 1990 and its contribution to the increase in emissions are less well documented than those of the fossil fuel industries. Emissions from power stations, coal mines, oil platforms and cars and ships are relatively easy to measure – now increasingly from space. Measuring the loss of carbon in soils, calculating the loss of rainforests and the net impacts on emissions, and working out the gap between 'biocrops' and what would otherwise have been done with the land are all technically more difficult. Yet we are not entirely ignorant of what we have done and the damage we have caused over the last 30 years.

The position in 2021

The last 30 years have witnessed an environmental disaster for the planet. We have not had the great transition away from fossil fuels that the architects of the UNFCCC promised. Things have never been better for many of the global oil, gas and coal companies. As for the renewables, they have played at best a very small bit-part, and almost all has come from hydro and bioenergy. Even here the impact of the dams and some of the biocrops has had some awful environmental consequences.

The last 30 years have also been the best for the agrichemical industries, but terrible for the soils and the peat bogs which store so much of our carbon. They have been great for the loggers, the palm oil plantation owners and cattle ranchers as the great rainforests have been cleared. The last 30 years have been beyond the travel industry's wildest dreams, with the great middle-class boom first in the developed countries, but now with China added too, all flying around the world for holidays. For shipping, the globalisation of the world economy, with China at its core, has vastly exceeded the expectations of 1990. George W. Bush was right: without China capping its emissions, little progress would be made. Almost every dimension of the climate disaster has China's name written all over it.

The 30 years have therefore been wasted from a climate perspective. Much more carbon has been emitted; the concentrations have marched ever upwards without so much as a blip, even allowing for the Covid-19 pandemic; the fossil fuel industries have boomed and natural sequestration has taken a nosedive.

How was this allowed to happen, when it was all supposed to be very different? That story is about the UN and Europe's efforts after 1990 – Kyoto, Paris, and the EU's carbon policies.

2

THE ROAD TO GLASGOW

The last 30 years were meant to turn out very differently. The UN was going to sort out climate change from the top down, by getting the nations of the world together and agreeing targets for each to limit its emissions. These targets had to add up to an overall cap on emissions, sufficient to stop climate change getting out of control. For reasons of expediency rather than any sense of an optimal climate, this was set at 2°C warming, and subsequently revised down to 1.5° as part of the Paris Agreement. Getting the world's nations together is what the UN does, and climate change gave it a whole new role. This is what the UNFCCC back in 1992, followed by the Kyoto Protocol in 2007, the Paris Agreement in 2015, and now Glasgow, are all about.

Thought of in this way, the climate change problem has been about how to get the recalcitrants like the US, those with targets well outside the overall envelope, and those who are not really trying to meet their targets, to change their behaviours. Put simply, the US had to get on board; Brazil and its neighbours have to stop destroying the rainforests; China has to take serious steps to deal with its coal and fossil fuel demand; and Africa and India have to get serious about both targets and actions on the ground. Fix all of these, and the problem will begin to get solved.

Except it hasn't and it won't. Why? There are two basic flaws: the design of almost any top-down treaty framework; and the way in which the UN and the parties have gone about it.[1] Even if the former were not fatal, the latter would at best produce a series of temporary fixes, limping from one 'agreement' to the next. It would keep the UN in the climate change business, but not do much about the underlying problems.

The fatal flaw

The ambition of the UN has been to create a carbon cartel. All countries are supposed to agree to limit their emissions by committing to caps, and these caps are together supposed to add up to reductions that will deliver a 2°C warming limit.

As a cartel, it suffers from all the problems of conventional fixes to markets – and some more.[2] Think of all the corporate villains in history who have tried to rig markets. They have used a series of bribes, extortion and even violence to get their way. Many have been simply criminal, with the Mafia perhaps the most notorious. Others have been encouraged by the State to bring economies of scale and hence greater efficiency. OPEC stands out as the international example.

What all have in common is the need to confront a basic incentive problem. In any cartel, each party has an incentive to encourage the others to restrict output (in this case, carbon) and benefit from the collective gains from a better outcome for all (less global warming). But each party also faces the costs of doing this. Hence the killer incentive: get others to do the hard graft, promise yourself to join in and then cheat. You get the benefits, but not the costs. It is what in economics is called the free-rider problem.

This perverse incentive is endemic in such agreements, and it is naive to pretend otherwise. All cartels and collusive agreements are designed with this incentive problem in mind, and, as we shall see, Kyoto and Paris are no exceptions. Most fail, and, to see why, let's think about how the incentive problem could be cracked.

The first (and for the UN an important) option is to appeal to common mutual universal human interests, beyond those of the countries and companies that form the parties to the agreement. In the case of the Mafia, it is an appeal to family loyalty, but it breaks down when there are other families in the game. That is why they keep killing each other. In the case of OPEC, it includes an appeal to Arab unity, and to the interests of the Arab peoples over and above the specific Arab nations. The trouble with this one is that Sunnis and Shia do not see themselves as part of a common cause, and OPEC needs Russia and now the US to at least tacitly fall into line, as witnessed in early 2020. The results for OPEC are obvious: every time the Saudis try to jack up the price of oil, everyone else starts increasing production, with the US and the North Sea as key examples. All OPEC efforts to create and sustain an oil cartel have run into difficulties, except in the very short run. For climate change and a carbon cartel, it is the medium and long term that count, and despite all the claims to the contrary, the UN's carbon cartel has fallen flat. That is why emissions keep going up.

The appeal to universal interests

The initial cheerleader for what became the Kyoto Protocol was the US, and Bill Clinton in particular. In the heady days after the collapse of the Soviet Union, the zeitgeist was captured by Francis Fukuyama's bestseller, *The End of History and the Last Man*.[3] The theme of the book is the recognition that, after trying almost everything else, including socialism, the rational enlightenment had produced its final end-product: liberal democracy. All nations would eventually converge on this model. Politics is a rational business, and to the extent that people are organised into nations, they would all come to share the democratic model, with markets to allocate most resources. Political dialogue would determine the success of this paradigm as it swept all other competing options away. Except it hasn't.

At the time, with the tearing-down of the Berlin Wall, and the resulting extension of freedom and democracy across Eastern Europe, the nightmare of the Cold War and the threat of nuclear annihilation gave way to a universalism which left less room for nation states. There would be diversity, but all within a rational framework. As people got richer, because liberal markets worked, they would lend their support to the liberal model. They would not, it was easily assumed, act parochially and nationalistically, as the old nationalisms of the past withered away. There would be no Donald Trumps, Vladimir Putins and Xi Jinpings, and no Marine Le Pens or Viktor Orbáns.

Behind this end-of-history thesis lay a deeper intellectual idea, one that was instrumental in the very creation of the UN. It was that rationalism would prevail, with a universal appreciation of the rights of all people, wherever and whenever they lived. It would find its expression in Nicholas Stern's *The Economics of Climate Change*.[4] Stern is a utilitarian who cannot see why we should discriminate between current and future people. It is easy to see why he thinks this. All humans are equal, and there is no moral reason to discriminate against any of them in favour of others.

The trouble with this particular argument and the rationalism on which it is based is that it is not how we humans behave. It is not what makes us human. We are partial, not universal, in our concerns. We do not transfer much of the GDP of developed countries to developing ones. We do not care about starving people in the Sahel as much as we care about our own children, our neighbours and our country. The UK struggles to persuade its voters to transfer just 1 per cent of GDP to those who could use it to much better effect than us – even to stay alive. Indeed, in 2020 it was announced that this would be cut back further. Climate change is partly a moral challenge, but it will not be solved by demanding conduct for which there is no evidence that we will entertain. Saints might, but not the bulk of the population who will be required to pay the costs.

In the heady triumphalism of the 1990s, and after the First Gulf War, it was not ridiculous to imagine that Russia would itself become a liberal democracy, and even join NATO and the EU. As for China, it was an easy assumption to make that capitalism and the liberalisation of its economy would usher in unstoppable demands for political liberty too. It might take longer for rogue states like North Korea, and even the religious regime in Iran, to fall into line. These countries in any event mostly depended on the protection of larger countries. As the government kingpins of Russia and China fell, they would find it hard to continue their authoritarianism.

Except it all turned out very differently. By 2019 there was Trump's 'America First'; Orbán's 'illiberal democracy'; Brexit; and Putin's determined grip on power in Russia – and especially on its fossil fuel industries. For anyone not carried away with the liberal democratic paradigm, the signs were there much earlier. Tiananmen Square, to anyone who watched the Chinese Communist Party's brutality in action, demonstrated that it was not going to lose its grip on power, and in due course Xi Jinping would consolidate power on a level parallel with Mao. The Uighur people, Hong Kong and Taiwan are the next targets.

Those who thought that politics would treat foreigners as of equal moral status to residents, and that religious strife and conflict would become a thing of the past, should explain why religion and ethnic origins remain so potent in Europe and the US, and especially explain the heightened tensions between Saudi Arabia's Sunnis and Iran's Shias and their allies. The old threat of serious global conflicts has not gone away. The US and China may well come to blows over Taiwan and the South China Sea, and Ukraine looks like a powder keg placed on the fissure between Russia and the EU. Climate change will impact on these nationalistic concerns, but it will not make them go away, and they will not be put aside for the greater good of tackling global warming.

Nationalism is the context within which global warming has to be tackled.

Appeals to a wider environmental conscience and for countries to act in mutual global interests – when the national self-interest incentive to free-ride remains – are not working, and they are not going to work. Such appeals make for great political speeches, and UN climate events are full of them. Hot air has so far been spectacularly ineffective, as witnessed by the continuing and relentless rise in carbon concentrations in the atmosphere.

Measuring emissions

Perhaps there is some other way around the free-rider problem? Take a look again at the conventional cartel dilemma. To fix output it has to be measured. Quotas have to be assigned to the parties; cheating has to be detectable, and the cheats have to be punished.

You can see where this is going. While we can measure the global concentration of carbon in the atmosphere, and while new satellite technology is getting better at seeing what is going on, it is work-in-progress to measure each country's emissions. Some bits are pretty easy, such as power stations and vehicle emissions. But what about all the non-fossil-fuel emissions? What about losses from the soil and peat bogs? What about all those forests being cut down? Soil, with four times the carbon of the atmosphere, really matters. Yet if you go to the Outer Hebrides in the UK, the landscape is littered with plastic bags full of peat cut for heating and even cooking. The peat bogs of Ireland still supply peat not only to the horticultural industry, but even to be burnt to generate power. The English Fens give off carbon on an industrial scale. Light a fire in your garden, or watch farmers burning crops outside Delhi. Who measures these contributions of Scotland, England, Ireland and India to carbon emissions? And who measures national carbon sequestration?

Preventing cheating

Measurement is slowly getting fixed. Preventing cheating is altogether harder, and in practice unenforceable for the major contributors. It requires threats and bribes which are credible: to impose enough pain and pay enough to offset the costs of action. A moment's reflection tells us that *forcing* the US, China, Russia, India, Brazil and many African countries to reduce their emissions is a non-starter. What exactly is Europe, for example, going to do? Trade sanctions against Brazil? Possibly. Direct action against China, Russia and the US? Unlikely, in a world of China First, Russia First and America First.

A recent example demonstrates how difficult this is. The EU negotiated a new trade deal with Latin America, the Mercosur Agreement.[5] It has been called the 'cows for cars' trade deal. As will be explored further later on, the 'cows' are partly coming from clearing Brazil's rainforests. Even if we could measure the consequences of the burning, and not just for emissions from the flames but also the loss of natural carbon sequestration, what exactly is Europe going to do about it? For the 'cars' represent jobs in export markets, and so punishment would be costly for both the EU and Brazil. Post-Covid-19, these export markets matter even more.

Brazil is a significant country, but not a great power. It is just possible to imagine it being bullied into submission. It could simply say that it accepts its global responsibilities and it is going to protect the rainforests. There is, however, no evidence that it is going to do so, and certainly not before a lot more damage (and emissions) has been caused. There is no evidence that China will stop buying its agricultural products.

Is there another way? Brazil thinks so: pay us to stop doing it. Its argument is simple and compelling: you, the developed world, put all the carbon up in the atmosphere as part of the process of transforming your economies so that you enjoy the standards of living to which we, the poorer Brazilians, aspire. Rather than threats and punishment, bribe us. If the developed countries do not pay up, then you will suffer from climate change, and you can't hit your 2°C target without us. The carbon cartel issue is therefore simple: transfer from the rich to the poorer countries whatever it costs to avoid the emissions that would otherwise enable the poorer countries to develop.

This way of thinking brings us to the old north–south divide, and the origins of the Brundtland sustainable development principle.[6] The Brundtland Commission was all about the distributional questions, and these are at the heart of the politics of the UN. Subsequently, the UN has tried to tack monetary transfers and funding arrangements onto climate change. So far it has failed, and for two reasons. The rich countries (and their electorates) don't want to pay up; and there is no credible way to make sure the money is spent on decarbonisation. Indeed, the incentives are even more perverse: Brazil could add to the extortion game by threatening to increase the damage unless the rich countries pay up. As we shall also see, this problem besets the carbon-offsetting opportunities: Brazil could get companies to pay to stop them cutting down more trees, rather than planting additional trees for natural sequestration.

When it comes to the great powers, paying up to poorer countries is in addition to the cost of doing the decarbonisation to hit the 2°C target. It is true that the great powers spend a lot on developing countries. The European powers have a long history of colonialism through conquest and occupation. The US has its own economic imperialism; Russia had its empire and now has its expansive military adventures. China has joined the 'great game', with its massive Belt and Road Initiative, and the acquisition of agricultural lands in Africa and Latin America. What the carbon cartel requires is that the great powers not only take measures themselves, but also pay large sums into a general fund *which they do not control*. Why would they do this?

What about the great powers' own actions and targets? There is a world of difference between talking the talk and walking the walk. So far, the EU, US, Russia and China have committed themselves to targets which are largely within their own interests. The EU, as we shall see, has been deindustrialising and its emissions were always going to fall anyway. The US is never going to sign up to a global treaty which gives the UN binding legal powers over it, without China signing up too. The Chinese Communist Party dictatorship is never going to do anything to jeopardise its power, and while it has a clear incentive to clean up the air pollution around its major cities, it needs its relative high economic growth trajectory to build its regional and global power and project its influence on the world – and even more so after Covid-19. It will, for expediency, adopt targets, but only those which it can meet anyway, or those for the distant future as with its 2060 carbon-neutral ambition, and that is indeed what has happened. The Russians have the opening-up of Siberia with more oil, more gas, and an opening north-east sea route, and the whole new global great game in the Arctic to play for.

There is one inevitable conclusion: the conditions for a credible carbon cartel are unlikely ever to be in place. But this has not stopped the UN from trying.

The UN's game

The UN is never going to solve this. The failure is not because of a lack of commitment by the signatories, or a lack of earnest moral intent. They mean well. The problem is altogether more fundamental. There is never likely to be a legally binding international treaty, with the teeth to enforce and punish those who fail to deliver on their commitments. The UN works (a bit) when the 'Big 5' permanent members of the Security Council – the US, China, the Russian Federation, the UK and France – force smaller countries into line, or when the problems are precise and well defined and it is in everyone's interests to fall into line. The UN has never resolved the Israel–Palestine dispute, and it has not driven nuclear disarmament. It is not going to crack the Brazilian rainforest crisis. It can be a very good mediator, and it can deploy peacekeepers, but unless all the Big 5 are on board, it struggles to get much traction. Nor does it have enough money to do the bribing.

The Big 5 do not agree on climate change, and they do not include emerging major polluters such as India, Brazil and African nations. They don't cover any of the world's great rainforests. The positions of the Big 5 are clear: Russia's elite has some interest in climate change actually happening in the short term, and, as noted, the US is never going to cede sovereignty. China talks a good talk but does not walk the walk. Of the two other smaller UN players, the UK is increasingly a bit-part player, especially after Brexit, and France is just not big enough and is in any event an outlier with its large nuclear energy system. The EU, the other really big bloc, is only indirectly at the table through France, but not Germany.

The UN Security Council is not and has never been the main act in global diplomacy, and has been a minor player of convenience in global crises. The action is bilaterally between the powers as they try to jockey among themselves for hegemony, and when important matters need cooperation it is typically the G7, G8 and G20 that provide the forums.

Notwithstanding the stony ground of geopolitics, the idea behind Kyoto was a grand one. Despite all the obvious cartel problems discussed above, and the interests of the big polluters, Kyoto was to be the first step in a process towards a global legally binding treaty, and that treaty would limit total global emissions to a level consistent with the IPCC's 2°C target. To do this would involve the parcelling-out of emissions reductions between countries. Since some were richer than others, and since the developed countries are responsible for most of the stock of carbon pumped into the atmosphere since the Industrial Revolution, the rich countries should take on targets, and the developing countries should take measures only. Amazingly, the architects thought it would actually work, and all this naive optimism is there in the speeches and papers.

The UN was encouraged by the now largely forgotten fact that the US was all in favour initially. Yet this proved a false dawn: it rapidly became apparent to Bill Clinton that Americans did not buy into this approach. He did not risk trying to ratify the Kyoto Protocol because there were no votes for it in the Senate. The objections were dressed up as concern about China not having to have targets, and hence creating an uneven playing field. But in reality, as the only superpower left, the US was not about to cede sovereignty to the UN, and not about to find itself paying developing countries to decarbonise. No US president since Clinton has really tried, and that is why the Copenhagen COP ended up with the Copenhagen Accord, and why Barack Obama could not sign up to a legally binding treaty at Paris. For all the very different styles and political outlooks, Clinton, George W. Bush, Obama and Trump have all taken the same US position. This is not about to change, despite Biden's rhetoric. US politics, as reflected in the Senate, remains sceptical of empowering others to direct its conduct. Biden advocates 'Made in America', whereas Trump wanted to 'Make America Great Again'.

Activists often present the US as the great villain of climate change, and understandably found it easy to demonise Trump, but less so Biden. That bit is not difficult – Trump does on occasion dispute the science and give oxygen to the idea that it is some kind of hoax, or worse, a conspiracy against the US. A more mature reflection on the other great players critical to reaching an agreement tells a more general story. None of them is going to sign up to international intervention in their economies, to cede sovereignty over their conduct. Neither Russia nor China would tolerate the sort of international supervisions which would be required to make tough limits stick. And then there is India and Africa, and those countries that are content to wait for others to act first.

The miracle solution

Despite such obvious structural flaws, and despite no progress on limiting the rise in emissions, the advocates of the Kyoto- and Paris-type processes have not let up. Maybe it has failed so far, and maybe lots of the optimistic rhetoric was misguided, but with one more heave will it work? The way the advocates put it now is that the difficulties are going to be overcome not by threats and bribes, but by the changes in the underlying costs of decarbonising. It is widely claimed that renewables are cost-competitive with fossil fuels already. Hence the problems dissolve: decarbonisation is no more expensive than sticking with fossil fuels. Wind and solar electricity generation, electric cars, hydrogen ships and planes, and biofuels and biomass are the future anyway.

This is the miracle solution. We can decarbonise and it won't cost us any more than not decarbonising. It's what the CCC would like us to believe. Even better, if we don't decarbonise we will end up with higher costs, because the fossil fuels will be more expensive than the wind turbines and solar panels that we will all be rushing to install. Decarbonisation is a win–win–win strategy: we get to mitigate climate change *and* we get cheaper energy too *and* it will increase economic growth. What's not to like?

The trouble is, it is unlikely to be true, at least for the next crucial 30 years. Eighteenth-century philosopher David Hume, the religious sceptic, wrote that, when confronted with someone claiming to have witnessed a miracle, there are two possible responses. The first is to believe this person; the second is to seek some other rational explanation. In the climate change case, the believers turn out to be activists and some European political leaders. But, as Hume pointed out, belief does not make something true.[7]

Think what the consequences of the miracle scenario would be. Just for a moment, assume that renewables are cheaper than fossil fuels, or will be in the very near future. There is no benefit for any country from not acting now. What could we expect? The most immediate change is relief from all the subsidies paid to renewables and for the other decarbonising measures. There is no need for intervention: the transition will happen anyway, and quickly as the great polluters will find themselves out-competed in world markets. It will be the Europeans and the UK who will be the shining economic successes of the future. Financial markets will clobber any fossil fuel producer as the value of their investments will nose-dive with the collapsing demand for fossil fuels. Russia, Saudi Arabia and the US, the three big producers, will face ruin.

There are many reasons why some might be delighted to see Russia and Saudi Arabia, with their authoritarian regimes, under severe economic pressure.[8] Some activists might also relish a setback on this scale for the US's new energy independence, built on cheap shale oil and gas. But the fact, as opposed to the desire or the belief, is that none of this is happening. It might, with new technologies from mid-century onwards, but not any time soon and certainly not soon enough. In fact, the news from the great fossil fuel producers is almost all good: Covid-19 aside as a temporary shock, the demand for oil, gas and coal keeps on going up, and all the mainstream forecasts and projections reinforce this message. As noted, the renewables (except hydro and some biomass and biofuels) have yet to make much of a dent. Just look at the market valuation at flotation placed on Saudi Aramco.

Nor are the activists campaigning outside parliament to end renewables subsidies. What they want is more, not less, money to compensate for their lack of competitiveness. The big 10 mbd oil producers – the US, Russia and Saudi Arabia – are not suffering from pursuing their narrow national interests. On the contrary, many of their problems stem from the falls in oil prices due to the superabundance of fossil fuels, now temporarily somewhat ameliorated.

The UN's role in this miraculous new world would wither away too. What is the point of all the theatre and spectacle and all the COPs and speeches (and all the air miles too) if the problem is going to solve itself anyway? If it is all already cost-competitive, what exactly is its role? It could campaign against the claimed fossil fuel subsidies, but such a fiscal matter is probably best handled through the International Monetary Fund and the World Bank. It could focus on development, but again this is a crowded field of international organisations. It might focus on the one bit that the miracle does not properly address – the destruction of natural capital and the capacity for natural carbon sequestration – and turn its efforts more to this natural capital agenda and biodiversity. Rightly, it already is. But in the round the miracle would push the UN back to its otherwise declining significance, reduce its staff, and further lower its profile.

The implication is pretty clear. The climate change question is: what to do, in the expectation that the top-down global agreement approach is not going to work, and even if it eventually does, not in time to avert significant climate change? And what to do, given the wall of economic growth – and therefore consumption – to come in China, India and Africa?

Europe takes the lead

If this is the question, it is one Europe chose to ignore, and is continuing to do so with some remarkably inefficient consequences. As Kyoto unfolded (it took until 2005 to come into force – 15 years after the 1990 baseline), it rapidly became a Europe-driven protocol.[9] The EU had three separate reasons for its enthusiasm. First, many Europeans took the challenge of climate change very seriously.

Second, the EU institutions and their leaders were looking for new ways to engage with their electorates and especially the young, and wanted to co-opt the burgeoning green movements into the core of mainstream political parties. Finally, they knew that the targets would be easy to achieve and there would be few costs. The reason for this last point was the baseline – 1990. This was when the Soviet Union collapsed, and hence the new Eastern European members would have steeply falling emissions from 1990. The Western EU member states were also well into their own structural deindustrialisation. However, because of poor domestic policies, Germany would struggle to meet its own 2020 target; for many it was a cruise. As it turned out, Germany did succeed in meeting its target – but only because the Covid-19 pandemic resulted in its biggest reductions in emissions in three decades.

The baseline of 1990 had one other advantage. Of all the countries whose emissions could be assumed to fall sharply from 1990, Russia stood out. The Soviet heavy industries collapsed with the regime that supported and subsidised them. Emissions plummeted, and indeed Russia itself went bankrupt in 1998. The EU needed Russia to sign up to Kyoto to make the Protocol operational – to pass the minimum requirements. Russia needed something else which the EU could gift: support for joining the World Trade Organization (WTO). The deal was done: the EU got the Kyoto Protocol to come into force; Russia was accepted as a member of the WTO; and nothing changed in Russia at all in respect of climate mitigation policies.

The need to have Russia on board was all the more pressing because others were dropping out. Two very new countries in emissions terms saw their domestic carbon production rise sharply. Canada bust the limits early on and formally left in 2012, and Japan struggled to comply, and dramatically so after the 2011 Fukushima Daiichi nuclear disaster triggered the closure of all its nuclear power stations, and a dash for coal and gas to fill the gap. Japan was second only to France in terms of the size and share of nuclear in its energy mix. With nuclear gone, it was hopeless to try to meet the Kyoto targets.

From within the European climate change political bubble, it was quite difficult to get a perspective on what was going on in the rest of the world. The crucial and disappointing fact is that no other country was seriously trying to control emissions, except for very specific reasons like air quality in cities, or because large hydro dams solved the burgeoning electricity demand challenge. China saw the Kyoto Protocol through the prism of its industrial strategy. If the Europeans were going to buy lots of wind turbines and solar panels, here was a new industry it could master. It duly did, on a large scale and with massive state subsidies. There were two consequences: it largely wiped out the European competitors; and it got to produce the steel and materials using its overwhelmingly coal-based heavy industries. A not inconsiderable side benefit was that the cost of solar panels fell.

The debacle at Copenhagen

Kyoto was always meant to be a step in the direction of an all-embracing international treaty, and the UN pushed on through the mechanisms of the COP to try to create such an agreement. The arrival of Obama in place of George W. Bush gave the process a new vigour (just as it would again when Biden replaced Trump), on the mistaken assumption that his rhetoric could be matched by votes in the US Senate. The stage was set for the Copenhagen COP in 2009.

On display were all the fundamental flaws in the process, replicated at every subsequent COP. It pitched the activists and campaigners against the politicians. It brought in thousands of people, many by air. The negotiations followed the usual UN pattern. As much as could be agreed in advance was, but this turned out to be very little. The negotiators were locked into late-night sessions picking over draft texts. The politicians turned up at the end, to take credit for the 'agreement' and to sort out the final issues. Except they didn't. So little of substance was agreed that by the time the key politicians turned up, the showcase event – Obama meeting Wen Jiabao – turned out to be the point at which it was recognised that there would be no Kyoto 2.

To save face, and avoid a clear rupture with the UN process, the ambitions were replaced by an 'accord'. This listed the voluntary pledges and offers made by those attending. And that was it: allowing the UN process to live another day, and have another go at getting a legally binding agreement. The pledges, in the unlikely event that they would actually be delivered, did not add up to the 2°C target requirements, and by a wide margin. The major world polluters were not prepared to do what it would take. Failure led to a doubling-down on the approach. One more heave, and the UN assumption was that they could get a legally binding agreement over the line.

The next big event was the Durban COP in 2011. The UN team arrived with the same old book. Durban was supposed to pave the way for a legally binding agreement, to be agreed at Paris in 2015. The delegates went through the same process. Just keeping the process alive was declared an achievement in itself. The summit concluded with an agreement to try to get an agreement in Paris.

The denouement at Paris

The gathering at Paris was supposed to bring Obama, coming to the end of his term in office, the Chinese leader and the rest of the key polluters together, finally, to sign up to a legally binding agreement, with legally binding targets.

The UN teams had spent the time since Durban flying from capital to capital, trying to cobble together an agreement. The EU pitched in early, and it was the EU that clearly led on ambition. It could, because Europe continued its deindustrialisation, whereas China and India were industrialising, and the US remained an industrial economy.

The outcome at Paris repeated what had gone before, with a few extra fixes.[10] Crucially, there would of course be no legally binding targets, as intended since Durban. Countries would again make their pledges, this time called 'nationally determined contributions' (NDCs), but if they missed them there would be no repercussions. There is no credible top-down enforcement mechanism, because no sovereign nation is going to allow others to impose their wishes on it. It would not be UN First, but rather continue to be China First, India First, Russia First and America First, as it always has been and probably always will be.

The UN team might have been relaxed about this if the perverse incentives could be ignored. What if it was economic to meet the pledges anyway? What if renewables and all the other measures, like protecting rainforests and stopping the depletion of the soils, were cheaper than fossil fuels, logging and intensive agriculture? What if the miracle solution outlined above is true? Lots of activists and some politicians repeatedly make this claim. Even if it cost a bit, it would not be much. Given that the IPCC had spelt out the dire warning of Armageddon just around the corner, and activists proclaimed that 'extinction' was coming, would not the world's big polluters decarbonise anyway?

This is where the pack of negotiating cards hit the floor. For if all this were true, if it was all economic, there would be no need for an international agreement, or protocol, or even a treaty. There would be no need for Paris or the continuing UN process at Glasgow.

Having failed to secure the legally binding targets, the UN's climate bureaucracy fell back on a second-best. Instead, the parties would be legally bound to come up with further non-legally binding

NDCs by 2020, and these would be compatible with the overall 2°C. It could then be claimed that there was a legally binding agreement, a triumph for Paris. If the current NDCs did not add up, at least there was a pledge to make sure they did in the future.

The trouble with this fig leaf is that that is all it is. How exactly was the UN to make sure that the new NDCs at future COPs – and Glasgow in particular – are going to be in line with the overall 2°C? What happens if, say, China offers a certain reduction, and let us for the moment imagine this is close to zero? Does everyone else have to take this as given, and reduce their emissions accordingly? This is the nature of the game, and of the enforced carbon cartel the UN is trying to create. Countries can make all sorts of pledges, and they could even be made 'legally binding', but what exactly is the UN going to do when (a) the pledge is not deemed adequate from one particular country; and (b) it is breached? Send in the UN peacekeepers to Washington or Beijing or New Delhi? Invade the Philippines? Of course not. That is obvious. And what is also obvious is that there are never going to be credible legally binding and credible legally enforceable national carbon targets.

Think about how Paris stands up against other global agreements. Take the WTO and trade agreements. The WTO is a set of rules, not a set of outcomes. It is ultimately about contracts, state aids and competition policy. It is a global version of the EU's competition regime and customs union. Members agree to abide by the rules. Except many do not. The recent trade wars between the US and China are all about the flagrant and very public violations of the rules of fair trade. China grants massive state aids to its big companies, insists on technology transfer, and lacks an independent judiciary to enforce contract and property law. It is China First. The WTO is very limited, as the UK found out when it contemplated a hard Brexit path. Trade is governed by trade agreements, and mainly on a bilateral basis. Further trade rounds after Doha have got nowhere, and the WTO is going through dark days in the face of America First and China First (Biden's 'Made in America' and Xi Jinping's self-reliance). Its perilous state reflects the general retreat from multilateralism.

Take another global threat – nuclear war. The UN could lead on arms-reduction negotiations, telling the parties what weapons they can have and inspecting them. It does some of this: UN inspectors were sent into Iraq, for example. Yet this role is confined to what the UN does best: inspection, expertise and information. The UN cannot tell Russia to get rid of its nuclear weapons, and it cannot regulate the nuclear arsenals of France, the UK, China, Israel, India, Pakistan and the US. Indeed, the very institutional structure of the UN is designed to prevent this sort of interference through its Security Council and the deeply embedded veto powers. All its members are nuclear powers.

Having failed to get the parties to come up with pledges that add up to 2°C, the Paris Agreement instigated the new target of 1.5°C. As long as it is an aspiration, the parties could happily agree that this is ideally what all the parties together would like. All of them know there is no chance of it happening. Recall that global temperatures have already gone up by nearly 1°C, and the UN's own IPCC scientists have spelt out the momentum that is climate change. Trying to halt the further temperature increases to 0.5°C is like trying to stop and turn around a supertanker. It takes time for the full impacts of what we have already done to the atmosphere to work through. It would take something like a major volcanic eruption such as that of Mount Tambora in 1815 to sufficiently darken the skies and throw the inbuilt warming momentum into sudden reverse. Yet even this was temporary as is the shock to global GDP from Covid-19. Nevertheless, 1.5°C is now the new target. Paris is not going to deliver it. Neither is Glasgow. We need something else.

3

GOING IT ALONE

If our limited attempts to control emissions in the last 30 years have achieved virtually nothing, and if Paris and Glasgow are not going to work, then we either give up or try something else. A number of countries, led by the UK, have decided that *unilateral* action is the best option. Whatever anyone else does, we should decarbonise anyway, in the hope that others follow.

But why would any individual country, faced with the evidence all around that key polluters are not playing ball, do this? Why go net zero *on your own*? In reply, unilateralists variously argue that unilateralism solves the incentive problem; that it creates exemplars for others to follow and hence is 'no regrets' for everyone; and that it is morally the right thing to do even if it does not solve the problem.

Unilateralism and the incentive problem

The most obvious starter question for the unilateralists is this: will it make any difference to global warming given the logic of the pursuit of national self-interests? Lots of answers spring to mind. One is that if only the US or China decided unilaterally to radically cut their emissions it would make a difference to global warming because their emissions are such a significant part of the whole. But the UK at 1–2 per cent? Unilateralism by big polluters is much more likely to be a useful strategy than by small countries if it is emissions themselves which are our focus.

Why would the US or China go completely unilateral? The US has understandably argued that its actions are conditioned by those of China. At the scale required to meet 2°C (or even 1.5°), it is not about to do this purely unilaterally, and has never been willing to cede significant competitive advantage by taking on world leadership in reducing its emissions. And as we saw in the previous chapter, China has never been willing to do a deal which meets these conditions. Not even when Obama got together with Wen Jiabao in Copenhagen back in 2009 was such a commitment forthcoming. The result was, as noted, the largely ineffective voluntary pledges that formed the Copenhagen Accord, outside the UNFCCC framework. Nothing has happened since to suggest that the US and China are going to agree about climate change to the full extent necessary, and they are not going genuinely unilateral any time soon. For all the hype around Biden, the gap between the rhetoric and the reality of what would need to happen remains vast. The Chinese leadership is not even playing the game.

That leaves the Europeans. Will this make a difference? The EU is roughly the same size in GDP terms as the US, with roughly the same level of emissions. Both represent around 16 per cent of the world economy, although the EU has a much bigger population: 515 million versus 330 million, so significantly lower emissions per head (see Table 1, page 30). Europe has been reducing its emissions unilaterally for a couple of decades, and the Kyoto Protocol targets ended up being European targets, some of which have been met. It now plans to go even further – to be net zero by 2050 and to have demanding interim 2030 targets too.[1]

The trouble with the Europeans pressing the others to fall into line is that Europe has not actually done much to reduce its carbon footprint, despite cutting emissions at home. And much of these cuts would have happened anyway. It is a story of deindustrialisation, and importing carbon emissions instead of producing them. Any other country considering unilateralism in response to what Europe has done, and especially if they look carefully at Germany and its *Energiewende* (the planned transition to a low-carbon, nuclear-free economy), could be forgiven for asking how exactly this gives them an incentive to follow. Germany does not plan to exit coal completely until 2038, athough it might now speed this up.

There is a way in which unilateralism can incentivise others, but this requires a more radical approach by the unilateralists, notably to start to properly price the carbon in their imports. This is the innocent-sounding 'carbon border adjustment', and we will explore how this works later on in the context of more general carbon pricing. But here the important point to note is that this is not so far what the Europeans have been up to, and in many respects the European unilateralism creates an incentive for others to do the opposite, by gaining further competitive advantages for their exports to Europe. Belatedly the European Commission has proposed to do something on this front, but not enough and not fast enough to close the gap.

Unilateralism and no regrets exemplars

The trade problem can in principle be fixed, and we will see how this can be done shortly. It is something the EU can and should do. But first, there is a separate argument which is used to justify unilateralism. It is a variant of one we have already met. Unilateralism, it is argued, will showcase examples of how decarbonisation can be achieved. World leaders will look to Europe, and to Germany and the UK in particular, as examples of a successful net zero strategy and its implementation, learn the lessons, and see how it can all be done. They will then want to follow these examples. This is very much the narrative at Glasgow.

Experiments have their virtues, and since the future is uncertain, and many of the decarbonisation issues are systemic in nature, watching how one country goes about it is likely to be very instructive. It would be especially so if the exemplar were lower-cost, leading to lower bills, greater economic competitiveness and, as a result, a political success too. The trouble is that the experiments so far have been more instructive in how *not* to do decarbonisation.

Some examples are better than others. The UK's unilateral net zero approach has at least not replicated Germany's biggest mistake. The UK has finally got round to phasing out coal, the first-best option for any decarbonising strategy. This not only cuts emissions fast and cheaply, but improves air quality, reduces water pollution, and limits the immense environmental damage from coal mining. It is a no regrets policy.

The UK's unilateralism goes one step further, trying to pick future global technological winners, and turning them into UK winners. Right now, very few of these low-carbon choices are obviously no regrets in the sense of being cost-competitive. Choosing renewable electricity remains more expensive than a mix with electricity generated from fossil fuels. Electric cars are more expensive, and organic food is more expensive than non-organic.

These things may be more expensive because of mispricing in the economy. With the correct – efficient – prices, perhaps they are competitive and hence no regrets? Once all the various other costs are properly taken into account, they might be cost-competitive, and hence the UK exemplar of going for net zero is one others should want to emulate?

Why? Because we do not price pollution, a cost which does not go away simply because we ignore it. An efficient economy, which would yield the highest rate of sustainable economic growth, is one where all these costs are properly taken into account. Polluters would pay, whether it be for carbon, for pesticides and fertiliser effluents, or for the particulates that get lodged in your lungs. They would pay for the metals that leak from mine water, and the costs of the clean-up after mines close. On the other side, landowners would get paid for the additional benefits from managing their land in an environmentally beneficial way, which would have the incidental benefit of sequestrating more carbon. Most ways of reducing carbon turn out to have other environmental benefits too, and reducing many of the other types of pollution turns out to reduce carbon emissions also and increase carbon sequestration.

A second feature of an efficient economy is that public goods are provided by the State and society, because they will not be provided by the market, and because they enhance the performance of the overall economy. Without the core public goods, the economy cannot function properly. The crucial ones here are infrastructures and R&D. Doing both properly is largely no regrets. Doing both badly, as at present, is a serious drag on the economy.

These public goods can be effectively provided only by public action, and the role of the State is to provide public money for these public goods. They can be paid for by taxes or levies on consumers. Add in the polluter-pays principle above (and its implication that any damage must be compensated for by net gain), and the beginning of a unilateral climate change mitigation strategy emerges that is efficient for any country to pursue because it is no regrets. The climate change problem is best seen as one that demands the polluter-pays principle and where public money for public goods policies are universally applied, and any one country can implement its own pollution pricing, invest in its network infrastructures, and fund its own R&D. An exemplar no regrets carbon target, and the corresponding policies, is of great value to the rest of the world. But as we shall see, it has to be the right target and the right policies.

Unilateralism and morality

If the free-rider incentive question stubbornly refuses to go away, if the unilateral strategies run into the problem of import substitution, and if unilateral policies are badly implemented (as we shall see in the German case), is there any other way in which the challenge of climate change can be configured, or are we all destined to sit back and watch the tragedy unfold? If a global agreement is not going to work, and if unilateralism can even be counterproductive, is there anything we can do?

This is where the moral dimension comes in. Even if our actions, on their own, might not make much difference, do we nevertheless have a moral obligation to stop making matters worse, and indeed to try to make some positive contribution? What can we each do to help save the planet?

The moral arguments come in many shapes and sizes. There is more to life than consequences, and while these matter, so too do our intentions and motivations. Utilitarians think that all we should be concerned about is utility, and in particular the sum total of utility. It is a consequentialist philosophy, adding up all the utility each person gets as a consequence of their actions to ensure the greatest happiness to the greatest number. It is not concerned with why people do what they do, but rather with what happens when they act. It is impartial: it does not matter which individuals, where, or in what time period, get the utility. Furthermore, since the marginal utility of money (the extra pleasure we get from each extra bit of money we have to spend) is assumed to decline as income goes up, transferring money from the rich to the poor raises aggregate utility.

Economists are typically utilitarians. It is, as discussed previously, the philosophy behind Stern's economic analysis of climate change. But it is not the only way to morally configure the climate change problem. While we cannot jump from *what is* to *what ought to be*, any moral position on climate change has to start with some of the inevitable limitations and constraints that human nature imposes.[2] The most egregious utilitarian assumption is the one that says morality should be completely impartial. We are miles away from this utopian idea. We care more about our family and friends than others; we care more about people in our own region and country; and we care more about people now than in some far distant future.

All of this matters because it determines how we should frame the climate change problem from a moral perspective. Take the individual unilateralist position. We could say that each of us should not make climate change worse, and that we have an obligation to make no net carbon emissions, so we are not personally responsible for any further climate change. We should do no harm.

The retort that this would make no difference to climate change is an obvious one, and as far as it goes it is true. Just as our individual votes are unlikely to change the outcome of an election, so our emissions are too trivial to impact on the global climate. Under a utilitarian consequentialist view, we should do nothing, and especially if doing something costs us some utility.

Many of us would not conclude that this utilitarian argument exonerates us from taking personal unilateralist action. We might regard polluting the environment as morally wrong, full stop. It might be that we conclude that living a good life involves care for others, regardless of the benefits to ourselves. Indeed, a cursory glance at how economies and societies work demonstrates that very few people pursue a narrow utility-based approach to others, and if we did, the economy and society would collapse. There would be no trust (since trust is always open to free-riding), and hence the economy and society would have no resilience, and be always open to short-term opportunism.[3]

Suppose then that we each want to do our bit, and we each would like to first limit and then eliminate our own net carbon emissions. Our personal climate change problem is how to effect this objective. We would start with our carbon footprint, and look at the multiple ways in which our lives generate carbon emissions.

To see how you can personally get to net zero, and more generally tread as lightly as possible on the planet whatever anyone else does, a good starting point is to turn to your own carbon diary, an approach you will recall from the introduction above.

Let's make the diary a bit more sophisticated. Try to work out how much of the stuff you record is made abroad, and especially in China. Look closely at the labels, and look behind the immediate stuff you buy and think of the underlying systems that are needed to make it. Think about where your smartphone, car components, clothes and furniture are made, and where your food comes from. Lots of this carbon is emitted in producing all your stuff outside the UK. Your personal diary will reflect the impact, however small, you have on the demand for carbon-intensive products made elsewhere. You should reduce your carbon footprint, irrespective of where the emissions are generated. You need to be net zero in your total carbon consumption.

To meet this moral requirement, you now need to construct for yourself the carbon diary *you would have had* had you met your moral goal. Go back to the beginning. Do you use only low-carbon sourced electricity? Do you buy your food locally? Do you get your food loose or in pre-packaged containers? Do you walk or cycle to work? Think about life under the Covid-19 lockdown and how your carbon consumption fell.

Some of these steps are easy, some are no regrets, and some are frankly pretty tough. But because you are a really moral person and you are determined to do your bit, you are prepared to make the effort. But what of your neighbours? What of the rest of the population? Maybe they are more selfish than you? Maybe they simply do not care about the environment? For them the issue is much simpler than for you. They are going to make the choices which are no regrets *for them* – choices which are worth making anyway. If walking to work improves their health, if eating less meat is good for them, then they may do so. Indeed, look around, and you will see that providing the information about the benefits of lifestyle changes does influence behaviour. Sometimes it needs a nudge. The NHS, for example, is now prescribing green options to improve mental and physical health.

Which brings us back to how to shape our individual choices and the no regrets approach, and to encourage others who might not be so altruistic. Imagine if the cost of carbon pollution was paid by the polluter, the infrastructure was provided and paid for, and the investments in R&D for new low-carbon technologies were all funded by the State. Now the prices you face when you go through the choices in your carbon diary are very different. Electricity generated from fossil fuels will be more expensive than the low-carbon options. Breakfast cereals in plastic packages and made from overseas ingredients will be more expensive than home-produced cereals like oats. Local food will be cheaper than imported produce. Palm oil will become an expensive ingredient.

The companies that supply you will face these costs in their inputs. They will have an incentive to choose lower-carbon alternatives – they might think twice about importing steel from China, for example.

We begin to see the light at the end of the carbon tunnel, and how unilateralism might work out. Done correctly, a whole lot of actions that would mitigate carbon emissions and encourage natural sequestration would be no regrets. Our actions could make a difference, especially if the economy properly reflected the costs of pollution and if the infrastructures and R&D were in place too. We could make our carbon diary a lot cleaner. Whether this would provide a good example for others to follow depends on whether this efficient economy is what we put in place, or whether we blunder from one costly mistake to another. As with the overall failure of the Kyoto and Paris architecture, many of the actual models followed in Europe, and especially in Germany, provide at best a negative example – how not to do it.

Unilateralism in Europe

A number of European countries have jumped the gun and declared unilateral carbon emissions reduction objectives. Europe put its 2020 targets into law, and has followed up with new regulation for 2050, and a target for 2030 as well as a net zero target for 2050.[4] The UK has gone furthest with its 2008 Climate Change Act and, in 2019, putting net zero into law.

The Europeans, having become the only players left in the Kyoto game, set about reducing carbon production on what was effectively a unilateral basis. The aim was to show leadership and provide the world with an example of how decarbonisation could be done without high cost. Germany went further than the rest and planned to develop decarbonising as an industrial project, not only contributing to its economic growth but also creating new national champions which would then have a competitive advantage on the world stage. German politicians believed in peak oil, and followed through its economic logic.

To do this, a new European climate package was eventually agreed in 2007 and translated into a set of unilateral directives in 2009. There would be 20 per cent renewables, a 20 per cent improvement in energy efficiency, and a 20 per cent reduction in total emissions – all by 2020, the Kyoto target date, and all against a 1990 baseline. All would add up to the politically convenient yet economically implausible magic number 20. By 2009, nearly two decades had passed since the baseline year, and so the targets had to be achieved in the 13 remaining years.

The architects of the 2020–20–20 targets intended to meet them, and the design of the directives reflected this. It should not have been difficult, and they made it even easier: the 20 per cent renewables included biomass, which would become 50 per cent of the renewables energy mix. Since there was a lot of biomass in Europe anyway, and since it would in time include the conversion of large coal stations to burn wood pellets, the 20 per cent was less than it seemed. Biomass was also, along with biofuels, of questionable value to decarbonisation. The 20 per cent energy efficiency target depended on what constituted an energy efficiency improvement, a very imprecise science, and eventually it would be more an aspiration than a binding commitment. Finally, the overall 20 per cent had as a backdrop the collapse of the Soviet Union in 1990 as a very fortuitous and convenient baseline. Taking all these factors into account, meeting the targets could probably have been largely achieved without any new policies at all.

The EU Renewable Energy Directive

The Renewable Energy Directive produced a lot of action across Europe.[5] It was a major victory for the wind and solar lobbies, and the rising demand for turbines and solar panels resulted in two developments: the cost of energy increased; and the imports of turbines and panels also increased.

The directive reserved one part of the market and kept it safe from exposure to competitive offers, and in particular to the other main strand of EU energy policy – the Internal Energy Market (IEM).[6] In the IEM, state aids and anti-competitive practices were ruled out in favour of the gradual unfolding of a more uniform, open, liberal and competitive European energy market. The idea was that the price differentials would fade away, and customers would be able to shop around for the cheapest electricity and gas supplies. There would, after a fight, be regulated third-party access to the networks. All this would help to address the perceived international competitiveness problem when compared with the US.

The IEM spelled death for the renewables. They were not cost-competitive and if customers shopped around they would all be out of the market. They needed state aids and they needed priority access to networks.[7] So while the IEM pushed towards competitive and open markets, the Renewable Energy Directive pointed in the exact opposite direction. The State would provide protected tariffs for reserved market shares, and force customers to pay through their use-of-system charges. There is no surprise in politicians pursuing two incompatible policies simultaneously. What is surprising is that they thought they could live in these two parallel worlds and escape the consequences.

This protected renewables market provided a great boon to lobbyists and vested interests. All they needed to do was capture the political process and then they would get their preferred technology subsidised and, as it turned out, substantial profits too. Capturing the political process proved easier where mainstream political parties needed to either capture the green votes or go into coalition with green parties (as in Germany).

Capture turned out to be remarkably successful. Whereas many European countries could rely on biomass, a number embraced the new frameworks and stepped up their unilateral commitments. Germany and Spain stand out among the larger EU member states, with Denmark at the smaller end, and the UK somewhere in the middle of the pack. Below we examine Germany's *Energiewende*, since it is on a scale and cost all of its own, yet was on course to miss even the 2020 carbon emissions target before the Covid-19 pandemic.

Spain has considerable advantages when it comes to renewables. Unlike Germany and other northern European member states, it has lots of sunshine, and hence very considerable solar potential. This it unleashed with generous subsidies, producing a boom. Spanish politicians were greatly encouraged by the incumbent energy utilities, and for a time forgot to take account of the costs to the energy customers. Eventually reality collided with the captured politicians. When the financial crisis hit at roughly the same time as the EU launched its 2020–20–20 package, it was clear that the wider population could not and would not pay the resulting bills. Spain was forced not just to slow down, but to retrospectively cut the generous subsidies it had offered.

Denmark projected itself as the great 'green' country, and especially as the home of wind power, both onshore and especially offshore. It promoted two companies, DONG and Vesta. DONG had grown as Denmark's offshore oil and gas company, and its offshore expertise was obviously helpful when it came to offshore wind. It was split into a renewables and a residual oil and gas business, and eventually the offshore bit would become Ørsted. As with much of Scandinavia, this was in the public sector, and hence the political capture was arguably easier.

Denmark rapidly developed its wind resources and then entered into other EU members states' markets. It could count on a really big carbon reduction. Yet all was not quite as presented. The reason Denmark could go heavily into intermittent wind was because it could import electricity from the north and spill its surplus wind to the south. To the north it could draw on hydro to balance its system, and there was the large German market to the south.

But what is really remarkable about Denmark is that, despite its small size and strategic location, it could not actually reduce its impact on climate change. It turned out that the increase in carbon intensity of its imports more than outweighed the decline in domestic carbon emissions. As a result, it continued to contribute to global warming.[8] The very high electricity prices in Denmark deterred domestic energy-intensive production, with imports taking up the slack. This last point about carbon imports is one that was reflected across Europe, as we shall see.

The Renewable Energy Directive applies to energy, and transport forms an important part of energy-related emissions. To address transport, the EU decided to use a mix of emissions standards and a requirement to use a given percentage of biofuels.[9] The result was a misguided dash for diesel and a push towards the production and import of biofuels.

Neither turned out well and the overall target had in due course to be watered down. Diesel turned out to create really big problems for air pollution and push a number of cities into violating air quality standards. It did not help that some of the big players in the car industry cheated in their laboratory emissions tests, producing a major industry scandal.

The biofuels turned out to be anything but green. Wheat production was diverted into creating them, and a lot was imported, produced from ethanol from sugar cane and palm oil. It is hard to argue that the transport initiatives did anything much to address climate change, and in fact they may well have been perverse. They did, however, boost farmers' incomes.

The lessons others learned from the EU's climate change package differed according to the interests. Wind and solar saw the EU package as a licence to print money, and they are quick to claim that the initial mouth-wateringly high subsidies were a good investment since they drove down costs, and now we all benefit from cheaper wind and solar. They never did a proper cost–benefit analysis (CBA): was all this spending strictly necessary for the benefits that accrued, and did they really mean that these cost reductions for solar at least would be Chinese and not German or European?

The lessons were rather different for other countries. Would their customers and voters stomach the high energy bills that resulted? Couldn't they sit back and watch, and then come later, gaining from the cost reductions that Europe had invested in? Unsurprisingly, no other country is rushing to replicate the EU's particular unilateral policies.

The EU Emissions Trading System

While the Renewable Energy Directive was designed as a way of protecting this type of technology from competition and from the IEM, at the same time the EU pursued a more competitive market-based general carbon policy. Although the European Commission toyed with a carbon tax back in 1991, it was captured by lobbying for a permits scheme by industry and the incumbent utilities. The EU Emissions Trading System (EU ETS) fixed the total amount of carbon that could be emitted by the power sector and large industrial plants, divided it up into permits, and then allowed these to trade to establish the EU ETS carbon price.[10]

The great advantage to the lobbyists of a permits scheme over a carbon tax is that the permits were given out free to the existing polluters (grandfathered), and hence the initial income effects stay with the polluters rather than going to the government. They were given the permits for free, and there was the advantage that *future* permits might also be given to traded sectors for free. The permits could be added to their balance sheets, and they provided a convenient barrier to other potential entrants into their markets.

A new carbon trading industry (and hence a new set of vested interests) emerged, profiting from carbon trading. The EU ETS also offered opportunities for corruption and VAT (value-added tax) fraud.[11] Instead of setting a uniform carbon price as a tax, and engineering its gradual rise over time, the European Commission landed the EU with a short-term, volatile carbon price, which in the 30 wasted years made almost no contribution at all to decarbonising.

It gets worse. The EU ETS played a significant role in maintaining the coal-burn and, at least in Germany's case, it flourished. For the greater the carbon reductions that were made by renewables, the easier the overall 20 per cent target was to meet. That meant that the room left in the overall carbon budget account was increased by the amount of carbon renewables saved. As a result, the carbon permit price could fall. In effect, the renewables were almost exactly offset by an increase in coal-burn, all courtesy of the EU ETS.

There have been belated attempts to fix these problems by politically manipulating the number of permits and therefore to jack up the carbon price. These, however, will have an impact beyond the first 30 years and, as we will see, the carbon price that may result could still be seriously inadequate.

Deindustrialisation and the UK

The really big story behind the unilateral policies and their failure to have an impact on the growth of the global concentration of carbon in the atmosphere lies in the long-run underlying economic fundamentals of Europe.

In the immediate postwar period, Europe was a manufacturing powerhouse, producing steel, petrochemicals, fertilisers, cars, lorries, tractors, trains, nuclear power stations, aircraft, and so on. This economic miracle was driven by Germany, France and Britain, but it stretched across the European continent into East Germany, Poland and the former Czechoslovakia.

The industrial crisis came in the 1970s as the European economic model began to break down. Sharp rises in energy and labour costs permanently damaged Europe's competitiveness and it would never fully recover. There began a gradual deindustrialisation, as other parts of the world picked up market share. The German economic miracle gave way to the Japanese and Korean economic miracles, and then to China's phenomenal economic expansion. The common feature of all these 'miracles' was the export of energy-intensive goods. From the late 1970s, Europe gradually ceased to be the first or even a major energy-intensive economy. In the UK, for example, whereas a 3 per cent growth in GDP meant a 7 per cent growth in electricity demand from 1945 to 1979, after 1980 this relationship broke down. This decoupling was the result of the changing industrial structure rather than climate change policies.

From a global climate change perspective, it does not much matter where the energy-intensive production takes place. If Europe's production goes down, this is only a global emissions win if it does not go up somewhere else. But it did and, worse from the climate perspective, it shifted dramatically over the last 30 years to much more polluting locations, especially China. (The Japanese economic miracle came to a shuddering halt in 1989.)

This matters for the impact of unilateral policies because what Europe did in the last 30 years was create an additional boost to a deindustrialising process already under way. The Renewable Energy Directive and the EU ETS increased the costs of doing business in Europe. What the EU achieved was to transfer energy and hence carbon-intensive production from Europe to China, and as a result make each unit of production more carbon-intensive than it would otherwise have been. It even encouraged the wind turbines and solar panels to be produced in China. All this offshored manufacturing was reimported into the EU.

However, none of this counted from the perspective of the EU's 2020–20–20 climate package. Or rather it did count on the positive side, without any offset for the imported carbon. Closing down energy-intensive production scored as an unambiguous plus, and carbon-intensive imports simply did not count at all.

The lesson others learned from all this is that a unilateral policy can actually speed up deindustrialisation. That is not what the US has had in mind, and China could sit back, talk the talk and encourage the Europeans to deindustrialise further, in the knowledge that it would be Chinese companies that would go on benefiting.

The UK's Climate Change Act

The UK went further than any other country by putting its unilateralism on a solid legal foundation with the Climate Change Act in 2008. The Act passed through parliament almost unanimously. Its centre piece was an 80 per cent carbon reduction target by 2050, which is legally binding. This has subsequently been increased to 100 per cent – net zero. To ensure that this would be more than wishful political ambition, an institutional architecture was put in place. The Climate Change Committee (CCC) is tasked with advising the government on progress and, in particular, with drawing up a series of rolling five-year carbon budgets, with three set at any time.

The carbon budgets represent the CCC's envelopes of emissions towards the 2050 target, and they are presented to parliament for approval. The government can reject these, but only if it comes up with alternative proposals which have the same effect.

As with the EU targets, the carbon-reduction target is a unilateral, UK-only emissions one, taking no account of imported carbon. The CCC made important assumptions early on, rendering it even more costly than it would otherwise have been. In particular, it assumed a linear line from the start through to 2050. This is surprising. Between now and 2050 there will be lots of technical progress, so it should get easier and cheaper to meet the target. The profile would then be less reduction now and more later.[12]

The linear approach is sometimes implicitly supported by an additional argument: that there are first-mover advantages from decarbonising on a fast track. However, as we will see in the more extreme example of Germany, there is in fact no evidence of the first-mover advantage. On the contrary, there are some distinct disadvantages, especially when added to its impact on carbon consumption. The knowledge, expertise and science are all public goods, and indeed the first-mover argument is undermined by the claim Germany now makes that going first provides for the transfer of such knowledge to other countries to assist them in decarbonising. If it is an implicit aim *not* to protect intellectual property (IP) in a first-mover strategy, then being a first mover cannot yield an economic advantage at home.

We are left with the argument that the linear decarbonisation is in essence aimed at the creation of the public goods in the IP. The empirical question is: has the UK, as a result of the linear approach, developed any IP which is of global value? The answer is that the only possible case is in offshore wind. The UK has some of the best sites – shallow water, near to the coast, and with good wind flows. Of all the technologies that the UK could develop of benefit to the world, this is probably the main one, followed by CCS – again because of shallow seas, but also the depleted existing wells and pipelines already in place, and carbon supplies ready to hand. Indeed, the North Sea is probably among the best global sites for both offshore wind and CCS. On hydrogen, there is no obvious competitive advantage, except where it is piggy-backed on CCS, using natural gas as the feedstock.

The UK has not yet tried CCS, and it does not need the linear CCC line to embrace both of these technologies. It could simply have an industrial policy to concentrate on them, and a more balanced approach would be a slightly smaller offshore programme and an early CCS one. Some offshore wind has demonstrated the technology and delivers logistics and other learning-by-doing gains.

A second assumption was that the agricultural sector was 'too difficult', and that the early carbon reductions should be concentrated almost exclusively on the power sector. This, based on a faulty economic analysis, is simply wrong. UK agriculture is of at best marginal economic value, and yet it covers 70 per cent of the land area. The value of its output per annum is around £9 billion, some 0.6 per cent of GDP. Of this, £3 billion came from direct subsidies via the EU's Common Agricultural Policy (CAP), and now post-Brexit will come from the UK Treasury. Then there is half-price 'red' diesel (a relative subsidy encouraging more farm-based carbon emissions), and exemption from business rates and inheritance tax. Furthermore, there is the exemption from the polluter-pays principle, leaving the pollution to damage biodiversity, and contaminate water, in turn raising costs for water companies and reducing biodiversity in water systems. In reality, the net value of UK agriculture is probably close to zero.[13]

Even this might be too rosy a picture. For the CAP protected UK and other EU farmers from global competition through its tariff wall. Most UK agriculture would not survive free trade, so in effect UK consumers pay higher prices to protect an uneconomic industry. That is the post-Brexit dilemma.

It follows that any loss in output from UK farming resulting from decarbonisation measures is likely to be trivial and lower than for any other area that policies have been directed towards. It is the lowest-cost option and a very low-hanging fruit. Better still, as we shall see in chapter 8, retaining more carbon in the soils, and increasing tree cover and hedges, has multiple non-carbon payoffs. Carbon in soils is a good proxy for biodiversity, and trees and hedges have benefits for water and flood management and enhance the mental and physical benefits that come from nature.

A unilateral carbon policy should therefore have started with agriculture, and then moved up the cost curve. Energy would still have to feature because, in a coming age of greater electricity demand, it will have to be decarbonised. In doing this, again the lowest-hanging fruit should be first, and that is coal. On this the UK unilateral policy has belatedly made the lowest-cost choice, and the stark contrast with Germany is apparent. The UK is not closing existing nuclear power stations; it is building at least one more twin reactor; and it decided to fast-track the closure of coal by 2025, using a unilateral carbon price to help deliver this outcome. Germany is fast-tracking the closure of all existing nuclear, building no more, and slow-tracking the exit from coal. There are clearly better and worse unilateral policies.

Germany's *Energiewende*

Of all the carbon myths, the presentation of Germany as the great green champion, and the greenest in Europe and indeed the world, is a triumph of spin over substance. From all the speeches and media coverage, you might think the Germans are the greenest Europeans, and with one of the strongest green parties in Europe, that their leadership would translate into carbon progress. The German reality is very different and an object lesson to the rest of the world of just how not to do it. As an exemplar, it was intended to show how decarbonisation could be done fast, cheaply and as part of a successful industrial strategy to create world-beating new renewables companies. The results have been almost the exact opposite. It has not decarbonised very much; it has been very expensive; and it has provided an industrial strategy boost for China, not Germany. No developing country would want to follow Germany's example.

What makes the German example brown instead of green are two key related factors. First, the *Energiewende* was as much a decision to exit nuclear as it was about renewables. Germany has a long history of anti-nuclear campaigns. Being on the front line of the Cold War, and with the backdrop of World War II, this is understandable. The Greens, having started as anti-nuclear pacifists before becoming green, had a nuclear exit as the first priority. The Red–Green coalition government from 2000 to 2004 was the first step in this exit, but it was repudiated by Angela Merkel's subsequent coalition, which eventually decided to extend the lives of the existing nuclear power stations. She then changed her mind after the 2011 Fukushima disaster, and in the face of a difficult regional election in Baden-Württemberg, suddenly announced a complete exit by 2022.

Whatever the domestic political rationale for this hasty decision (and it was very hasty), the result was that Germany would increasingly rely on imported nuclear-generated electricity from its neighbour, France, and would face whatever nuclear risks its neighbours' nuclear power stations might present. Chernobyl had affected all of Europe. What followed was a need to generate the electricity from something else, and to do this quickly and, given the key role of the nuclear plants in the south, to build lots of high-voltage capacity from the wind in the north.

Germany's unstated strategy was to rely more on coal. This was highly polluting, damaging to human health, and in direct conflict with the 2020 targets. It is hard to think of a policy that could be worse from a climate change perspective. Not only would the existing coal capacity be kept going, but 13 GW of new coal would be added. Only Japan made a similar disastrous move, but then it had had the Fukushima disaster at home and had shut down more than 50 nuclear reactors.

The coal that Germany now burned over and above what it would have with the nuclear stations resulted not only in higher carbon emissions, but lots of particulates and sulphur too. Those living near the lignite coal fields, and those miners who worked in them, had their health impaired. Premature deaths are part and parcel of coal mining and coal burning, and relative to nuclear it has proved much more harmful to human health. It was a 'multiple regrets' policy.

As 2020 approached, Germany was still struggling to get rid of its most polluting coal industry. It has now agreed to phase out coal by 2038, hence allowing itself another 20 years of coal burning.[14]

Yet it has been even worse. Not only has Germany gone from nuclear to coal, but also from gas to coal. Instead of using gas as a bridge fuel to cheaply displace the coal – and at half the emissions – Germany has done the opposite. In this, it has been aided by the EU ETS, which set a low carbon price, gave gas little cost advantage over coal and, as noted, enabled the renewables to be offset by coal.

It is true that the Germans embarked on a grand solar project, with very high subsidies, and lots of wind too. With its relatively dark northern skies, Germany put in place the largest subsidies for solar in the world. This was so great as to form a significant part of China's massive solar panel production expansion, in the process creating a bubble of over-production globally. The result has been multiple bankruptcies and the wiping-out of German solar companies. Costs fell as a result of Chinese mass production, and this has indeed been a global benefit, but it was not the intention. Rather, the aim was to produce world-leading German companies. There is none.

Adding all this together produced some of the highest electricity prices in the world. This completed the picture: Germany had shown it would struggle to meet its own 2020 target (in the absence of the pandemic impacts); it had increased associated pollution from coal; it had not created winning German companies; and it had demonstrated to the world how to make electricity really expensive. All of this is on a carbon production basis. On carbon consumption, it could at least point to the maintenance of a higher domestic manufacturing base, by shielding the corporate sector from some of the costs, but even this is now giving way. Energy-intensive German steel and petrochemical companies are no longer investing in Germany.

Belatedly, Germany is having another go at reducing emissions. It is implementing a carbon price on transport fuels and adding another €50 billion of subsidies. In doing so, it was aiming to break its balanced budget (before Covid-19 came along), so future generations will have to pay.

A place for unilateralism

Unilateralism has a strong moral foundation. It is what we should do to tread lightly on the planet, but it should be done properly. If you want to have no net effect on global warming, there is only one way to do it. You have to focus on your carbon consumption, to be net zero in consumption and not just production. That applies to countries too. The lesson for the EU, the UK and Germany is clear: look to the carbon border adjustments to make sure that it is really net zero consumption that is being pursued. This is ultimately the only way of being sure about not contributing to further global warming. If it genuinely is net zero, it is radically more demanding than any of the policies discussed in this chapter. It is what we must do if we really want to tackle climate change.

The merit of pursuing a unilateral carbon consumption net zero target is that it might also influence the decisions of other countries. A carbon border price, as we shall see, has the rather neat property that it incentivises others who want to export to us to impose their own carbon tax. This, rather than the UN top-down Paris approach, might actually broaden and deepen the coalition of the willing.

PART TWO

The Net Zero Economy

4

LIVING WITHIN OUR ENVIRONMENTAL MEANS

Since a global deal is not going to be enough anytime soon, and since we can't afford another 30 years like the last, unilateralism offers the main practical route forwards, particularly if it is based on no regrets policies, rather than the chronically inefficient route Germany and Europe have been taking so far.

The climate change problem, like other environmental problems, is the consequence of an unsustainable consumption path. If this is generalised, then it has a much bigger implication: that we are all on an unsustainable growth path. The pursuit of ever-higher GDP, and a belief that growth is maximised by ever-expanding aggregate demand, regardless of the composition of that growth, is what has brought us to the current situation.

The starting point of reconstructing a sustainable economic model and, in the process, cracking the climate change problem, is with us and with finding a way to tread more lightly on the planet. When you look at your carbon diary, you will immediately identify the huge gap between what you are doing now and what you need to do. It requires quite radical changes. Simply boosting consumption to boost GDP is the road to climate and environmental ruin. Sadly, it is the path we are on, and it is one that most central banks and most governments are determined to keep us on, for fear of a recession.

What would a sustainable economy look like? The answer is that it would be one where all the costs of pollution are internalised; where the key public goods that the market will not deliver unaided are provided; and where any environmental damage requires compensation. These are the three general principles of a sustainable economy: polluter pays; public money for public goods; and net environmental gain. Our policies on climate change, set out in subsequent chapters, would basically be the implementation of these principles into the fabric of our economy.

The merit of an economy that conforms to these conditions is that it would be efficient, and therefore it would be able to deliver the highest rate of economic growth consistent with that growth being sustainable. It is the best we can do, while protecting the only planet we have. Once the principles are in place, the question that then follows is: what sort of economy, and what balance between the State and the market, would best deliver this sustainable outcome?

The first principle for a sustainable economy: The polluter pays

Climate change is caused by all sorts of pollution. These are part of the costs of doing business and the costs of our consumption. Without a price, the costs are ignored. Because they are ignored, the amount of pollution will be excessive: there is no incentive to reduce it. That is true of carbon, methane and all the other greenhouse gases, and the principal reason why too much of all of them is emitted.

Note the word 'excessive'. The optimal quantity of pollution is rarely zero. To see why this is so, consider a world in which there is zero pollution. It would be a world with very few humans. There would be few if any crops, and we would live only on those renewable bits of natural capital which are abundant and well in excess of any threshold below which they could not renew themselves. We humans are always changing our world, and there is no prospect that this will end, unless of course there is some cataclysm. There is no wild to rewild back to, no primitive and pristine state of nature to which we can aspire.[1]

The point is to make our impacts in such ways as to leave the natural world in at least as good a state as we inherited it. We should tread lightly, but nevertheless still tread. This is not and can never be a zero-pollution world. But we have already done enough harm.

In the context of carbon, there is in any event no prospect that we will cease all carbon emissions; nor should we. The aim is to stabilise the amount of carbon in the atmosphere. That is a stock, and it is the volume of this stock relative to the other gases that influences the climate. The natural processes of the earth are removing carbon continuously. That is what plants and trees and oceans do. The aim is to fine-tune those emissions to keep the concentration from going up further. Greenhouse gases are a 'good thing' in the right concentrations in the atmosphere, and it is the balance that counts. We can help to get this roughly right by not only reducing our carbon consumption, but also helping natural sequestration.

By introducing a price for pollution, everyone is incentivised to stop emitting where it is easier to do so, leaving the residual emissions to those areas where the value of the carbon is highest, and hence where it is most expensive to avoid the pollution.

A carbon price helps with those who are not persuaded to act morally themselves by creating a selfish incentive for them to switch from carbon-intensive goods, not because they necessarily care but rather just because these become more expensive. How to introduce a carbon price is the subject of the next chapter.

An efficient economy is one where all pollution costs, and not just those of carbon, are internalised by being priced. Imagine if all pollution was priced, and that the polluter-pays principle was generalised across all sources of pollution. The economy would be *radically* different. Put another way, our current economy is staggeringly inefficient. As with the expansion of many economies over the course of the twentieth century, our economy has been based on the extraction of non-renewable minerals (including but not limited to fossil fuels) and the devastation of renewable natural capital.[2]

To see how inefficient this has been, consider what the legacy is for the generations to come in this century. Because the polluters did not pay, the pollution has been and continues to be excessive. The costs can be seen all around us. The atmosphere is full of CO_2 and other greenhouse gases; the rivers are polluted with industrial effluent, fertiliser and intensive agricultural runoff; and the seas are acidified. Some of the damage is irreversible, including the large-scale extinction of species that this economy has caused. All of it is inefficient.

Who is the polluter who should pay? It is easy to jump to the conclusion that the large corporations are responsible for the legacy of pollution and that still being added; that it is the giant oil and petrochemical companies, the agrichemical firms, and even the farmers, that as the guilty parties should pay.

No doubt there is much that these companies can and should do, but they are not the ultimate polluters. It is you and me – and therefore the polluter who should ultimately pay is you and me. We buy the stuff the companies make for us. We buy the petrol and diesel, the flights, the plastic-wrapped products, and the cardboard boxes that our internet purchases come in. We buy the clothes which are often barely worn before being discarded. It is all the stuff in your carbon diary.

Looked at like this, the consequences of the polluter-pays

principle are more challenging. It is fine when 'they' are paying, but when the bills come home to us, the pain is unavoidable. Making the supermarket 'responsible' for the plastic wrapping means that its prices are going to go up. So is the cost of all the steel and cement and fertilisers which make up key inputs into the stuff we buy. If you go on a train, you should be paying for the carbon emissions from the steel that goes into the rails. If you use a building, maybe a gym, it is made of cement and steel and plastics. And so on.

You might think that the companies can afford to pay and do not need to pass on their extra pollution costs to you and me from the application of the polluter-pays principle. For example, what about all those profits and dividends? Couldn't they be used instead to reduce pollution? Tax business rather than individuals is a familiar cry. They will suffer, not us.

This is just an illusion. Profits are the reward for the deployment of capital and labour. They are not a luxury. Remove the profits and the good or service just isn't produced, unless by the State, in which case the taxpayer absorbs the risks. The profits come from producing stuff we want and are willing and able to pay for. It all comes back to us again.

Applying the polluter-pays principle means that the relative prices of all the goods and services we buy change. Now certain things in the supermarket trolley will be more expensive. They will tend to be those items that involve a lot of manufacturing such as processed food, single-use plastic items, and all those chemicals and detergents for cleaning your home. Some items will now be cheaper. They will tend to be food produced closer to home, and organic food in particular. Avocados and other exotic foods will now reflect all those air miles and the costs of clearing rainforests to grow them. High-tech indoor farming close to home, using light-emitting diode (LED) lighting from energy generated from solar panels, hydroponics and controlled pest-free environments, will be the way year-round strawberries and beans are produced, rather than flying them in from Peru and Kenya. You will now think quite carefully about the choice between taking a holiday abroad and one at home, since the cost of the air ticket will be much, much higher. Domestic tourism will boom.

There will be a radical change in what goes into your shopping trolley. That is all to the good: it is how we shift towards a more sustainable, and efficient, economy. But there is a downside. Your shopping trolley will also be more expensive. Some things will be cheaper than others, but few prices will go down, and most will go up, some by a lot. Why? Because the pollution cost gets added to the other costs, rather than ignored, and that pushes up prices.

Higher prices mean that you are going to be worse off. Instead of enjoying the benefits of all those goods and services without taking account of the pollution consequences, now you will have to. Higher prices mean your income does not go so far. You will consume less as a result, and that is what you have to do to make sure the next generation does not get your rubbish in the atmosphere, on land and in the oceans, and instead gets a set of natural capital assets at least as good as those which you inherited. Another way of putting this is that because we are living beyond our, and the planet's, means, cutbacks are inevitable.

The politics of the lowering consumption that the polluter-pays principle implies is going to be tough. Nobody likes to face up to their environmental footprint. Politicians will have to stop promising a painless transition to a sustainable future, and economists will have to stop telling us that decarbonisation is going to be just a huge economic opportunity, all gain and no pain. It is not. The CCC is going to have to stop pretending that it will cost 1 per cent of GDP, or even nothing. Whether the political processes, democratic or otherwise, can deliver an enlightened self-interest which involves quite a lot of pain is the issue upon which the future of our climate, and planet, depends. If they cannot, then we fry.

Getting from our high-pollution, low-efficiency economy to a sustainable one will require a transition. Although there are some merits in a short, very sharp shock, going immediately from the current low prices to prices which properly account for pollution, the medicine might kill the patient. Those who, for example, advocate getting to net zero by 2025 or 2030 should think hard about just how profound this shock would be, and the risk that it would paralyse economies, throwing people out of work and inducing real hardship. It might make the economic consequences of Covid-19 look like a picnic. A sharp and immediate fall in consumption would lead to a knock-on further fall in demand. You will buy less in the supermarket, and that in turn will lead to lower demand by the supermarket for all the things on the shelf. As the cutbacks kick in, the economy will contract. The obvious risk is that this becomes a downward cycle, with mass unemployment. The voters are likely to revolt.

There is an example to hand. When the Soviet Union collapsed, it was obvious that all the artificially low, subsidised prices would have to be adjusted upwards. Some economists said it should be medicine administered immediately. Just unleash the market forces. There will be pain, but rapid recovery and growth.[3] Thirty years on and we are still dealing with the consequences of a shock that ended up with Putin and other authoritarian leaders in several of the former Soviet countries.

A reduction in consumption is a reduction in aggregate demand. That indeed is part of the point, reducing consumption to a sustainable level. Keynesians, and almost all governments, believe that the way to economic prosperity is to *increase* consumption, and especially in the short term. The shock of a sudden net zero transition would, according to this view, tip us all into a severe recession. Keynesians believe that borrowing for spending is generally a 'good thing', and that we should not overly worry about passing on the debt to the next generation because the extra spending will create growth and hence pay for itself. It was the logic behind Donald

Trump's tax cuts and spending increases, Joe Biden's massive recovery spending, and Boris Johnson's new-fangled 'cakeism': the Johnsonian approach to having the spending *and* greater prosperity *and* decarbonising.[4]

A critical distinction comes into play here. Aggregate demand is made up of both consumption and investment. Investment in the 'right things' holds up demand and keeps people employed. When it comes to climate change and indeed all the other environmental challenges, the scope for investment is very great. The transition is very much about rebasing and switching away from consumption towards investment. That investment should enhance the natural capital passed on to the next generation. It is about new greener infrastructures, natural sequestration, fibre and communications and R&D – all essential for decarbonisation.

There is an analogy here with World War II and the transition from a peacetime to a wartime economy. Governments crushed consumption with high taxation and rationing, and then rerouted the tax revenues to investment in war machinery and the associated wartime infrastructure. Now, the revenues from pollution taxes could be rerouted to green infrastructure spending and R&D. This is a switch, whereas many advocates of a 'green new deal' want to maintain consumption, paying for investment through borrowing, and not through higher levels of saving. They want aggregate consumption to go up to increase aggregate demand *and* to decarbonise, and get future people to pay back the debt. What climate change demands is to reduce consumption to a sustainable level and use the savings generated to pay for investment, as in a wartime economy.

The second principle for a sustainable economy: Public money for public goods

That extra investment will come in handy for the key public goods, many of which have significant impacts on carbon emissions and climate change. Public goods are, in economics terms, defined as non-rival and non-excludable. They are public because these technical characteristics undermine private incentives to produce them. If something is non-rival it means that you and I can consume it at the same time without additional cost. Think of listening to a radio programme as a classic example.

The economic problem of public goods, that users cannot be excluded, is compounded by the non-rivalry, which means that the marginal cost of providing to an additional person is zero. There is no cost to extra people consuming it. For example, once an infrastructure for charging electric cars is in place, the marginal cost of using the charging point is usually zero. If the electricity provided through it comes from wind or solar, then its marginal cost is zero too. The economic trouble with this is obvious: nobody has any incentive to invest in the charging network since they are not going to get paid unless they can exclude some people, and those excluded because they cannot pay could have benefited without imposing extra costs. Without government intervention imposing excludability or recovering the costs, there will be under-provision. At the limit, there will be none. It is hardly surprising that so much of today's infrastructure in many developed countries, like the US, Germany and the UK, is in such poor shape.

This has a particular and big impact on climate change because lots of the emissions (and the solutions to decarbonisation) involve infrastructure networks which have some public goods characteristics – the railways, the roads, the electricity systems, water pipes and communications. Networks have the characteristic of near-zero marginal costs, and either they need some form of exclusion mechanism or the State has to pay for them through taxation.

It is not just providing the network capacity to meet demand. Core infrastructure and utility networks are so critical that there needs to be extra capacity to ensure the security of their supply. More power stations are needed than strictly necessary to meet mean expected demand, for the reason that the unexpected might happen. Our roads and railways have this security margin too (even if it does not always seem like it). Again, there is no incentive to provide this margin.

It is therefore hardly surprising that these networks are subject to significant intervention and are often state-owned. They are mainly nationally planned, and most have universal service obligations (USOs), requiring operators to meet not just the demand of those willing and able to pay whatever price is imposed on users, but also of those citizens who cannot.

This all matters because the climate change problem requires transport and energy to decarbonise. In turn, the energy and transport systems cannot function without a digital enabling communications system, and that in turn needs a fibre and broadband infrastructure to support these decarbonising networks. Recent arguments about whether broadband and fibre should be free throw into sharp relief the issues at stake.

Consider some of the requirements of decarbonisation. Air transport, at least short-haul, will need to be replaced by trains, and we probably need high-speed networks (although not necessarily HS2 as currently configured). The roads will need to be converted for smart, digital and autonomous electric vehicles, and perhaps hydrogen infrastructure too. The electricity networks will need to enable the electric transport charging and storage, as well as handle intermittency and decentralised electricity generation, and help with decarbonising heat. This is an enormous investment requirement, and on a scale that matches (and probably exceeds) the World War II investments discussed above. It will require the savings to match the investment, and these will have to come from a combination of higher user charges and taxation. All therefore ultimately come from a switch from consumption to investment within the envelope of sustainable economic demand. The problem for many developed economies, and especially the UK, is that we don't currently save much at all (pandemic notwithstanding) – not even enough to cover our pensions and university tuition fees, or to pay back the national debts that are mounting up. The delivery of these infrastructures is considered in chapter 6.

The second major public good which will be under-provided by the market is R&D. To see why this is a public good, consider the invention of a new idea, say the worldwide web, or the new material graphene. Once invented, it is non-rival, and the idea is non-excludable. No one is going to invest in something unless they can get their money back. For this there needs to be protection of the new idea, by a patent, for example. Yet such protection to create exclusion limits the benefits to the few who pay, even though many could benefit without extra cost, because it is non-rival.

For these reasons, the promotion and support for R&D is a major function of government. Governments support and finance education (the transmission of ideas) and primary research. Much of this is 'blue skies', without obvious or immediate benefit. So much has been discovered as a by-product of this sort of primary and general research.

This is the general case for supporting R&D. Climate change adds a great urgency. Humans have developed the ideas behind the carbon economy, and in the narrow sense of GDP the carbon economy has been a transformational success in the last 200 years, when almost all the economic output for all human history has been concentrated. Ideas have enabled the world's population to grow to 7 billion – ideas about childhood diseases, antibiotics, and a host of medical breakthroughs that reduce infant mortality and have greatly extended the human lifespan. Ideas drive the development of Covid-19 vaccines.

The task now is one which demands yet more revolutionary ideas: to create a new sustainable economy, one that radically gets away from fossil fuels. Since almost all of our economies are dependent on fossil fuels, this is obviously an immense challenge.

It will happen only with significant government support for the research. This, and the supporting infrastructures, are what public money for public goods as a second key principle is all about. The R&D will in turn create the technologies which increase sustainable economic growth.

The third principle for a sustainable economy: Net environmental gain

The third principle is a particular subset of the polluter-pays principle, and of special relevance to climate change. The net gain principle requires that any damage done must be compensated for by more than the expected damage, on the precautionary principle given that there is always uncertainty about whether the net gains will actually materialise as anticipated and produce the predicted results.

The net gain principle itself comes in several versions. The narrow one is that specific damage to natural capital is addressed. Property developers are, for example, required to ensure net biodiversity gain. In a net zero carbon world, emissions would have to be compensated for by emissions reductions elsewhere, net carbon gain.[5]

Given that there will be carbon emissions for a very long time and practically for ever, and that the optimal level of pollution is not zero, net gain is about offsetting these residual emissions. This can be at the economy, the company or the individual level.

Suppose you decide that, notwithstanding the pollution caused and your personal responsibility for those emissions, and notwithstanding that you could go on holiday in the UK, and by train or electric car rather than on a domestic flight, you decide to take that foreign holiday and fly, just like millions of your fellow citizens. Calculating the full emissions caused (and not just by the flight itself), you could consider buying an offset, paying someone else to do things like planting trees to soak up the carbon. This may alleviate your guilt about that holiday somewhat.

But should it? Let's assume that your trees are actually planted. Are they really *additional* trees that would not otherwise have been planted but for the offset you purchased? Taking the example discussed earlier, suppose your offsetting company 'buys' tree credits from the Brazilian government. Suppose they say: 'We are currently cutting down the rainforest at one hectare per minute. If you pay us enough, we will "save" the forest from felling on a per hectare basis.' Is this really compensating for your emissions? If you and many others don't pay, there will be lots more emissions and a lot less carbon sequestration by the rainforest, so yes, the net carbon effect of your offset is indeed to offset your emissions.

But now recall the incentives on the Brazilian government, which has an implicit (and even at times explicit) policy of allowing the cutting-down of the Amazon rainforest for economic development. That development is typically for cattle, themselves displaced by the growing of sugar cane and soya, partly to make ethanol for use as a biofuel. The new Mercosur trade deal with the EU would open up the beef export market. Would it not now be in the interests of Brazil to announce a more rapid felling policy, in order to harvest the carbon offsets and increase beef production for the new markets?

This is an obvious perverse incentive. It leads to a more profound question. Are there *any* offsets which could reduce the net emissions caused by your choice to fly? Think of any measure that could be taken and ask whether it is an offsetting investment that should be made anyway. Are there any really *additional* carbon offsetting measures that would not have been made anyway? Chapter 7 considers the details of how sequestration works in practice, and the choice between natural sequestration and CCS. It is where agriculture and land use come very much into play.

Assuming for the moment that it is legitimate to buy an offset, how much net gain over and above the mean expected damage should be required? The reason why extra gain is necessary is because of the risk that the offsetting investment might not work as well, as noted above. But there is a further reason: the scope for corruption and institutional failure. Take the example of planting trees. You pay a sum of money to the offsetting company. It then uses the money to plant the trees. But what happens next? Who cares for the trees, maintains them, replaces those that die, deals with diseases, and generally carries out the capital maintenance? Assuming the company does not simply run off with your money, it could plant the trees, pay out its profits and close down, thereby avoiding the costs of the maintenance. How exactly can a private business guarantee the long-term success of your offset? For this reason, you might want to consider giving the money to an environmental charity, but even here there can be institutional failure.

This is why there needs to be net *gain*, and probably quite a big net gain. It is not the marginal cost of the carbon damage that should shape the price of your offset, but that plus quite a lot more.

Delivering a sustainable economy

A sustainable economy is one where all three principles apply: the polluter pays; the key infrastructure and research public goods are provided; and there is genuine net environmental gain. This is efficient: not to price pollution leaves costs out, which do not go away just because they are ignored; not to provide public goods means that lots of necessary building blocks for the economy, and for managing the transition away from carbon, are not in place, and ideas and technologies necessary to beat climate change do not arise; and not to enforce net gain means an economy where excess damage happens.

A sustainable economy has to meet intergenerational equity too.[6] It is not just pollution *now* that counts against efficiency, but also pollution dumped on the next generation. To address this generational point, the sustainable economy needs to meet one further condition: that it passes on to the next generation a set of assets at least as good as those it inherited. This in turn requires an intergenerational balance sheet which sets out the liabilities we are passing on – the carbon in the atmosphere, and the other sources of pollution – and requires us to create sufficient extra environmental assets to offset these, so that natural capital overall does not decline. If all three principles are regularly applied, and if investment is paid out of savings, then this intergenerational condition would be met.

When it comes to carbon, this generation is almost inevitably going to fail under this intergenerational criterion. The parts per million of carbon that we pass on will almost certainly be much higher than the level we inherited. So we need an additional compensation to make up for the fact that the next generation is going to get all those costs from climate change, while we benefit from not paying for the pollution we are causing. This could be achieved by creating a sovereign wealth fund, building up the funds that future generations will need in order to deal with the mess we will have left them with. They should in effect borrow from us through our bequeathing these monies set aside, rather than us borrow from them by building up debt and expecting them to pay. Without this, current consumption is too high, and the economy is pursuing an unsustainable growth path. It is a requirement that goes well beyond putting aside some of the benefits of, say, depleting North Sea oil, used for our immediate gratification in lower taxes. It is notable that the UK has not even managed to do this.

Future generations should also benefit from any sustainable economic growth which takes place from now until their time comes. Some environmentalists claim that there can be no more economic growth at all, and hence the future generation will not be better off in this respect.[7] The argument is that the earth has finite resources and hence the population runs up against natural constraints. It is not just that we *should* not have economic growth, but that we *will* not. It is the argument advanced by Thomas Malthus, and later by Paul Ehrlich and the Club of Rome.[8]

This is far too pessimistic. There is (sadly) no chance of us running out of fossil fuels, nor of many of the other main minerals that are crucial to the economy. But, more importantly, there is one resource which we continually invent and do not run out of. It is *ideas*, and these lead to new technologies which are passed down the generations. This is what gave us smartphones, laptops and new materials. It is ideas and the technological possibilities they create which have enabled humans to rise from their natural origins to become such an extraordinary species, capable of reading, writing, computing and inventing one technology after another. Whether for good or ill, the speed of technological change is accelerating. It will need to: the existing technologies cannot crack the climate change problem on their own and feed the world. We will need to open up the light spectrum, invent more new materials, and peer into the genetic box to transform food production.

That is what will drive sustainable economic growth. The environmentalists are right to argue that current GDP growth cannot be sustained, but that is because GDP measures the wrong (and often highly polluting) things.[9] There needs to be a rebasing, to a sustainable consumption level, and then growth can continue as ideas, science and technology increase human possibilities.

What is the best way to deliver this sustainable economy? Should it be driven by the State and an ever-larger role for that State, or should it be left to markets? What kind of economic model will do the job best?[10] Is socialism or market liberalism the right lens?

These questions are what the new politics of climate change home in on – with very different narratives on the left and the right, and with the left making most of the running. The young want radical change and rebel against the perceived intergenerational inequity of dumping on them the legacy of abysmal environmental stewardship. The older socialists see climate change through the prism of their critique of capitalism. Pollution is just a symptom of the capitalist system, which needs to be brought down and replaced by state planning, state investment and state delivery. A confluence of these different environmental-political strands has taken place, and not all of it is going to help achieve an efficient decarbonisation.

Many climate activists have worked out that in the sort of democracy we have, and with the market-based form of capitalism, there is probably never going to be a majority for the sorts of action they claim is needed to replace capitalism, or at least not until it is too late. We will start to fry before we vote for the necessary changes they demand. Recognising that the electorate – us – is not going to come on side in time, they have the intellectual honesty to recognise that they will need to change the democratic model in order to change the economic system.

It is here that things get scary. The activists would have us sort these things out through 'citizen assemblies'.[11] These assemblies might be just enhanced focus groups, randomly selected, to inform public debate. There is nothing new or revolutionary here. But they may go further and give the assemblies a more decisive role. It doesn't take much imagination to recognise that the results from these assemblies will depend on who sits on them, and some activists assume they will be people like them, who share their analyses and conclusions. If, on the other hand, they are a representative sample of the electorate, they will simply reflect the electorate, and that might not produce the outcomes the activists want. They might just vote for populist policies, including lower energy and fuel bills. Worse still, they may not care so much about climate change. They may pay more attention to politicians like Trump and Bolsonaro and the populists.

The democracy problem these citizen assemblies throw up is

countered with the argument that the assembly members will be making their decisions on the basis of 'the science' and 'future generations'. 'Listen to the scientists' is a regular refrain. It is the old problem of democracy in a modern guise: how much education should people have before they can vote? Should educated people have more votes than the uneducated?[12]

The obvious problem is that the educated – and the scientists – are not always the 'good guys'. Although the core theory of global warming is pretty much established and hard to refute, almost everything else in climate science is open to challenge. The decisions have to be made on the balance of probabilities, not certainties. Scientists are humans too, and in the name of acting on what many see as an almost existential threat, few also highlight any of the benefits of climate change, and many point to particular weather events as evidence of climate change without real understanding. Climate change can become a shared scientific belief, a paradigm, and is not always subject to the conventional caveats that should apply to any scientific advice.

Think back over the last 200 years and consider some of the things that have been done in the name of science. Scientists gave us DDT (the chemical compound Dichlorodiphenyltrichloroethane), and thereby facilitated the destruction of much biodiversity as agriculture has been chemicalised. Scientists also gave us the atomic bomb and the modern war machines that hold all of us in a state of potential annihilation through nuclear war. It was not so long ago that environmental activists were deeply suspicious of science.[13] Now scientists and activists agree, so the scepticism falls away.

The point here is not to challenge the science *per se*, but rather to say that there is a good reason for holding us all to account through the ballot box. The challenge is to get all of us to vote for change, not to replace the less educated and less knowledgeable with 'assemblies' of like-minded people. This is a pragmatic as well as a principled point: the policies needed are going to be painful in the short-to-medium term, and if there is not wide democratic consent then there will be revolt, as witnessed among many populist movements, not least the *Gilets Jaunes* in France whose first demand was the reduction of fuel taxes. Educate and inform, yes; replace and decide with the chosen climate elite, no. It is not only undemocratic. It will not work.

With citizens' assemblies shaping the decisions, the left argues that the State needs to control the means of production. It has long argued for the State to own the commanding heights of the economy. For much of the twentieth century it did, and in much of Europe it still does. This gives governments direct control over investment, and here the left's argument has further reinforcement. It can channel savings into investment. That saving comes from taxation, what is in effect forced savings. Governments during World War II imposed very high tax rates and then used the money raised to pay for the war. The same model was then applied during peacetime. Taxes remained high, and investment in housing, transport, energy, health and education was directed by the State.[14] It is a model that China has emulated over the last 30 years, while Japan and Germany once directed their high private savings through banks to these enterprises.

If it is true that decarbonisation is going to cost a lot, and in particular require a lot of investment, then raising taxes and directing the investment through state-owned industries is a very coherent approach. It can in theory get the job done. The onus, it might be argued, is on those who resist both the higher taxes and the nationalisations to show how it could otherwise be achieved.

The left needs one further measure to make sure that its ownership of production translates into the outcomes it decides we should have. It needs to rule out competition, to make sure that we do not make the 'wrong' choices. This is what was done after 1945 in the great UK nationalisations (although it had already been started during the war). Competition was regarded then as generally a bad thing, distorting choices towards the short term and undermining investments. It was made illegal in energy (coal, electricity and gas), and in much of transport too. In agriculture, farmers were protected from competition, and in due course were guaranteed prices for their outputs.

Although it became fashionable to dismiss this model from the 1980s onwards, in many respects it worked. The Central Electricity Generating Board (CEGB) built the power stations to meet the burgeoning demand for electricity. British Gas successfully (and efficiently) converted from town gas to natural gas and helped the North Sea gas industry develop in a planned and coordinated way.[15] The motorways were built.

Those on the left who recognise that the net zero transition is going to be expensive naturally focus on the position of the poor, and propose distributional policies to help them out. Free home insulation for the less well-off is an obvious example of the sorts of policies the left advocates. But piecemeal distributional measures and specific subsidies only scratch the surface. They need to go deeper, and this involves taxation. For the left the carbon tax might raise the money, but it is seen as regressive (although this is far from uncontroversial). The obvious answer for the left is to use progressive income and wealth taxation, and then regulation rather than taxes to fix the pollution.

It is not difficult to see what a hard sell to the electorate this is going to be, if taxes are to pay for the massive investment needed. It takes wars to get people to vote for substantial tax hikes, and successive left governments have found it very difficult to introduce even modest wealth taxes. This is where the honest activities get overtaken by less honest ones. In the US and some European 'green new deals', the tax issue is avoided by advocating borrowing instead, playing the classic Keynesian card. New national investment banks and national transformation funds are going to be tasked with borrowing to finance the investments. It is going to be pay-when-delivered, rather than pay-as-you-go (the model under the nationalised industries in the postwar period). Some go even further and argue for a 'green QE' (quantitative easing), printing the money to pay for the investments.[16]

The trouble with borrowing rather than raising taxes is twofold. First, it assumes that the multiplier effects of all this spending will increase economic growth and hence pay for itself. It relies on the borrowing increasing spending, multiplied through the economy. It is the sort of argument the right in the US advanced in advocating unfunded tax cuts under Ronald Reagan, George W. Bush and Trump. Biden's version is unfunded spending, including just giving people money to spend on a historically unprecedented scale.

This is a conventional Keynesian GDP growth approach, and there are lots of reasons to be sceptical about its likely success and the impact of the debts that result. If the general problem with GDP growth is that consumption is unsustainably high, it is hard to see how boosting consumption through borrowing is going to make the economy greener. The remarkable thing is that many environmentalists actually believe this.

The second problem comes with who gets to inherit the debt. It is the next generation, the young, who will get the pollution we are causing now and that we have already caused, and the debts to pay for clearing it up. This hardly fits with intergenerational equity, and it looks worryingly as if the current older generation of polluters is going to get away with it.

The honest activists are right: the current generation should face up to the consequences of its actions, and it should have to pay compensation to the next for the horribly polluted legacy it is bequeathing them. This, however, is an altogether harder message to sell. Yet it is correct, and does not suffer from the politically slippery Keynesian notion that we can have our cake (net zero) and eat it too (carrying on the current consumption path).

What markets can and cannot do

Among the many difficulties with the left's approach, in addition to that of selling it to the electorate, is its chronic inefficiency. In replacing the role of markets, the State has to do a great deal for which its competence is very limited. While the left is correct in that unfettered *laissez-faire* is not going to solve the climate change problem, replacing markets will not either. What is needed is a middle ground – combining a model with the State doing what the markets cannot, and the market then sorting out the resulting allocation of resources within the framework the State provides.

To understand this, let's take a hard look at the right, and why its solutions are just not credible. We can then sort out the appropriate economic borders of the State, and the relative roles of the State and the market in getting to net zero.

The right starts with a simple and elegant theory of markets. It turns out to be an argument by assumption. Assume that there is a perfect set of property rights, perfect information, profit-maximising firms and utility-maximising individuals, no monopolies and no market power. Assume all this and the market will arrive at a perfectly competitive equilibrium allocation of resources. Any change would make at least one person worse off.[17]

It is an ideal type, rather like the ideal type of perfect state planning. Our world is miles away from either of these. Uncertainty is pervasive, especially about the future within which climate change needs to be cracked. We are what Kant famously referred to as 'the crooked timber of humanity', an idea that Isaiah Berlin developed.[18] What makes us human is doubt, uncertainty and a whole host of motives and desires which cannot be captured by utility. We live in a world with imperfect property rights, imperfect information, and all sorts of market and human failings. Climate change is one of the consequences.

The idea that, left to the markets, climate change could be cracked is nonsense. Carbon is a negative externality and climate change a global public bad. R&D is a public good. Imperfect information leaves us resting on the balance of probabilities. All these market failures are large and feature centrally in any net zero path. But the State is also riddled with failures – government failures. The question is whether it is possible to reconcile the two – use the State where it has, on balance, the advantage; and use the market when it has been adjusted to deal with the market failures.

Rather than replacing the market, and losing its inherent incentives and the efficiencies that follow, a pragmatic climate change policy involves the three interventions set out in this book to correct the key market failure: making polluters pay; providing the public funding for public goods; and ensuring that damage is compensated for by requiring net environmental gain. This is what leads to a sustainable economic consumption and growth path.

Notice what the sustainable growth model does not require: deficit spending, Keynesian style. It requires that the polluters pay, and that is us, in this generation, since we cannot do much about those who put the carbon up in the atmosphere in the past. It has got nothing to do with the design and level of monetary or fiscal policy. It does not require state-ownership, although it does require a powerful state to price and regulate the pollution (and hence incorporate the costs), to provide for the public goods, and to enforce compensation through compulsory offsetting. Without the State carrying out these functions, and funding the public goods, markets are going to be hopeless at mitigating climate change.

Notice too that all of this is unilateral: we can do it ourselves, provided the price of pollution is generalised to include imports too. It also does not require an international treaty or agreement like Paris. It allows us to get on with decarbonising and stopping our contribution to climate change. It remains to fill in the details – how to price carbon, how to build the future infrastructures, and how to make sure that carbon is sequestrated. The next three chapters deal with each of these in turn, to put flesh on this sustainable consumption and growth path.

5

THE PRICE OF CARBON

If carbon consumption is the problem, there are two ways of reducing it. The first is to persuade people and to emphasise their moral duties. This has merits, but is unlikely to be easy, and will take a long time. There is the free-rider problem to consider (why should you do anything if no one else does, and even if they do, why not take that free ride on their efforts?), and there is the sad fact that not many of us are sufficiently moral when it comes to climate change. We prefer our consumer society, even if it is saturated with carbon. We fly, we drive sport utility vehicles (SUVs). Even at the height of World War II, rationing had to be compulsory because moral persuasion was not enough.

The second option is to make us, the polluters, pay a carbon price.[1] It is the first part of a unilateral strategy. It solves two problems: it is collective, in that all of us are caught by it and so there cannot be any free-riding; and it does not rely on moral persuasion. Doing the right thing by choice is a bonus.

Price, applied comprehensively, is the way to ensure that the polluter pays and, as we shall see, it has one other huge advantage. Done properly, it incentivises other countries to follow suit. It turns out to be a much better way of taking the bottom-up approach to encouraging international action – much better than Paris, and all the jaw-jaw which has yet to make a dent in the rising concentrations of carbon in the atmosphere.

It is hard to imagine any coherent decarbonisation strategy without a price of carbon at its core. If something is not priced, the pollution it causes is not taken seriously in everyday decisions by governments, companies and individuals. If the price is zero, or too low to represent the damage being done, there will be over-consumption of the things within which carbon is embedded. There is, and this is one of the main reasons why.

Arguing that it is all too politically difficult to confront people with the consequences of their pollution is really just saying that decarbonisation is not possible because people want to carry on living beyond their means. Without a carbon price, decarbonisation will almost certainly cost more, and hence be even more painful.

There is an obvious ethical objection to the carbon price: it represents a licence to pollute. This needs to be dealt with first, before turning to the practicalities: whether to fix prices or quantities; what price to set; how to adjust it over time; the domain of the tax; the point at which the tax is levied; who does the fixing; and what to do with the money.

The ethics of pricing carbon

Some environmentalists worry that putting a price on carbon is a licence to pollute.[2] They worry that polluters – especially rich ones – would be able to get away with their pollution. They also raise the question of the behavioural consequences, in that a carbon tax may lead people to believe that in paying it they have discharged their moral responsibilities for causing climate change. That flight abroad for an extra holiday does not come with a big dose of guilt if a carbon tax is paid.

These are serious objections, and they are about much more than just carbon. The use of pricing generally for these sorts of environmental challenges raises concerns of principle. Many economists compound the moral problem with crude CBA, making the mistake of equating costs and benefits with values. To say that something costs something is not the same as saying it is worth that cost, and saying what the benefits of something are is not the same as saying that the calculated benefits equate to its moral worth.

Economists get into this mess because they view people only through the lens of consumers, starting with the premise that people are rational, utility-maximising machines. It is especially troubling when this sort of utilitarianism gets enmeshed in arguments about the environment. The time horizons are long; decisions can be irreversible (such as frying the planet and extinguishing species); and future generations' utility can only be guessed at. Such calculations have no room for citizens, and no room for other ethical principles, such as the duty of stewardship – to leave the environment in as good or better shape for future generations.

Fortunately, neither utilitarianism nor the economists' narrow path are necessary to make the case for a carbon price. All that is needed is to recognise that there is a cost involved in polluting the atmosphere, and that those responsible for causing this cost should be confronted with its consequences. We can take a separate view about the benefits, and indeed the standard economist's toolkit – CBA – plays almost no role in identifying targets like the 2°C ceiling and the net zero goal. This makes the task analytically much easier: it is not CBA, but rather just CA (cost analysis). The benefits can be left to a mix of the science, identifying what is feasible, and the modelling of the temperature projections for various scenarios (such as the work done by the IPCC), and to moral philosophers who can guide us on our environmental responsibilities to future generations. These are not marginal decisions, a bit of utility here versus a bit of utility there, but rather system ones.

Put it the other way around: what happens if there is no price, and hence the polluter-pays principle is not applied? How do we decarbonise? The practical answer is that we could use regulation instead. We could use emissions performance standards, set targets for the proportion of electricity from different energy sources, mandate the closure of coal and the fitting of CCS to gas-fired power stations, and set dates for bans on petrol and diesel cars and gas home-heating boilers, and so on.

We do quite a lot of this already, and some of it is complementary and necessary alongside a carbon price. Some decisions cannot be taken purely on the basis of the carbon price. An obvious example is investment in, and closure of, nuclear power stations. These raise very long-term issues about waste and societal–political questions about the technological risks and nuclear energy's relation to military development of nuclear weapons. The choice of public goods, such as the provision of greener infrastructures and R&D, cannot be solved with a carbon price, although it can help even in these cases.

Price is not a perfect instrument, but it is one major and necessary part of a decarbonisation strategy. It is necessary, even if clearly not sufficient. The flaws pale into insignificance when compared with those of regulation. Regulation, especially in choosing technologies, is much more open to lobbying by the vested interests which plague the climate change scene. In energy, the economic rents from the subsidies and the mandating of particular 'winners' can be substantial. The result is a host of 'special cases'. Farmers get subsidised ('red') diesel and have so far escaped the consequences of stripping carbon from the soils. Landowners have been allowed to degrade peat bogs. The biomass industry has managed to get itself called a 'renewable' and hence get under the subsidy umbrella of the EU Renewable Energy Directive. With so many subsidies and exemptions, the result is that some of the most expensive technologies get picked first, raising the costs of decarbonisation, and increasing the political resistance to the higher cost of energy that is implied, since someone has to pay. This has not, however, stopped the Treasury from supporting the CCC's 1 per cent GDP cost of decarbonisation. The Treasury, like the CCC, simply assumes that there will be no government failures – only market failures.[3]

In the regulation world, what the interested parties need is an effective lobby group. Renewables UK has emerged to play such a role and has been remarkably successful, rivalled only by the National Farmers' Union (NFU) in its subsidised domain. They exist to pursue their members' private interests, not the public interest. The more powerful they are, the greater the distortions away from the public interest to the private interests of their members. Regulation plus lobbyists equals more expensive decarbonisation.

Regulation of consumption is particularly difficult. People do not like to be told what they cannot buy, and they hate rationing. Rationing might work in a war, in the context of a completely planned economy where the price mechanism has been all but eliminated. While there is a lot of food safety and health and safety regulation, top-down instructions tend to backfire. We have not managed to ban cigarettes, and indeed when it comes to recreational drugs, regulation has not worked well. The Americans tried prohibition, with limited success. Labelling is about as good as it gets – valuable and worthwhile, but hardly sufficient for the task ahead.

The case for a carbon price as a necessary accompaniment survives all of these challenges. It is about cost, not value. It is separate from the issue of the targets and their definition, and is generally much better than regulation, although regulation nevertheless has a *general* supporting role to play. It nudges you towards low-carbon products, without telling you what you can and cannot buy. You don't have to be a utilitarian or a CBA mainstream economist to support a carbon price.

Prices or quantities: the EU ETS versus carbon taxes

How should the price be set? There are two options: it can be a direct carbon tax; or it can be indirectly 'discovered' through a market in carbon permits. The first is straightforward: someone picks a price (we consider who should do this later on), and then it is raised as a tax, just like any other commodity tax. The second option starts with the target, works out how much carbon can be emitted consistent with that target, and then parcels up the allowed emissions under this target into a set of permits. The permits are then either handed out (typically on the basis of grandfathering past emissions) or auctioned to the highest bidder. Trading them facilitates the allocation of the permits to those who value the polluting activities most, and away from those for whom it is easier and cheaper just to cut the emissions. The trades set the price of carbon.

So far so good, and you might think that permits and taxes come down to the same answer. The tax can be varied so that the target is achieved, or the target can be turned into quantity-based permits, which will trade at a price similar to the alternative carbon tax.[4] In both cases, the end goal is achieved – in theory.[5]

It turns out very differently in practice, which is why the major polluters campaigned so hard for a permits scheme rather than a tax. Why? Because if the permits are grandfathered (handed out to the polluters at the start of the scheme), the polluters get a free right to pollute and, unlike the tax case, do not have to pay up to the government. They keep the money – what in economics is called the income effect of the carbon price. In theory and as noted, if the permits are all auctioned, the bid price will be the capitalised value of what the tax would have been. But in practice, permits are regulated property rights, and having an auction over a continuous time period is wide open to lobbying and political manipulation.

Governments prefer the money now, and spend it, and companies argue about the 'special circumstances' in their industries. Then there are the financial institutions which look to make money from the trading, something which is a deadweight loss compared with the tax.

With these lobbying and political influences inevitably at play, it is hardly surprising that, as discussed in chapter 3, the world's major trading scheme, the EU ETS, has so far been a failure. It is partial in its coverage, as the lobbyists campaign for exemptions and special treatment; it has high administrative costs; is open to criminal corruption; and produces a price which has been both volatile and so far pathetically low for most of its history. It has nevertheless helped to protect coal, and in particular in Germany, since more renewables reduce emissions and therefore leave more room for coal. We met this aspect of the *Energiewende* earlier.

The consequences have been so bad that the scheme has been repeatedly doctored to try to jack up the price. A host of interventions have been applied: to manipulate the number of permits; to deal with 'banking' of permits between periods; and to play with the scope. This all requires a lot of regulation. The European Commission has promised to learn the lessons, do it better and get the price up. But then it has made similar promises in the past. In any event, if it has the political intent to manipulate the market to produce the 'right' price, why not simply set the price and avoid the hassles, costs and corruption of the EU ETS? Why not just set the carbon price? The answer comes back to all those lobbyists yet again: they make money out of the EU ETS in a way that they would find much more difficult if confronted by a uniform carbon tax. Post-Brexit, the lobbyists have won again: the UK has opted to implement its own ETS – the UK ETS – rather than a carbon tax.[6]

Given that the case for a carbon tax is so powerful, why hasn't it happened? Here again the politics resurfaces. The politicians do not want to confront voters with the costs of their pollution, and a carbon tax does this very much 'in your face'. Recall the earlier discussion about the myth that decarbonisation will not cost much, if anything, and the easy political offer of painless decarbonisation as a path to economic growth. Once the truth is out, that decarbonisation is costly and will force us to live within and not beyond our environmental means, voters get higher bills. One of the quickest ways to get demonstrators out on the street is to raise fuel prices – the *Gilets Jaunes* protests in France being a recent example, joining other diverse examples stretching from Hungary to Chile to Iran. Even raising fuel duty has proved politically too difficult in the UK, and Biden has ruled it out in the US.

The lesson to take from this political reluctance to introduce a carbon tax at an appropriate level is that voters need to be told the truth. Pretending that decarbonisation is cheap, by hiding behind the subsidies and permits, has the great political merit of disguise. But it cannot abolish the costs, and the indirect measures are almost always more costly and inefficient, and open to political manipulation. In the end, the costs will be revealed, and the real political question is whether to tell the voters, or carry on the pretence. What cannot be avoided are the consequences. If there is not an appropriate carbon price, then decarbonisation at scale will almost certainly not happen. The politically easy bit is grandstanding about net zero and the great targets that are being set. But targets without the means to achieve them, and preferably as efficiently as possible, are just hot air.

What price to set

What is the appropriate carbon price? There are two possible answers. The first is the one favoured by economists. It is derived from calculations of the marginal cost of carbon abatement and the marginal value of the benefits from that marginal reduction, ignoring the system characteristics. There have been lots of attempts to calculate this social cost of carbon and, unsurprisingly, they yield a variety of answers.[7]

For the moment, accept this approach and consider the assumptions that need to be made. What exactly is the cost curve of marginal reductions of carbon? What are the cheapest options for reducing a tonne of carbon?

The numbers that result from all sorts of engineering and other studies range very widely, from less than £10 per tonne to more than £200. Unsurprisingly, too often the studies produce the answers that the different interests want. As for the benefits of avoiding emissions, these depend on how much damage is assumed to be caused by the emissions. Whereas emissions anywhere have much the same impact on the climate, the damage costs to the economy of climate change vary considerably by location. They are bad news for those in the tropics, but less bad for developed economies that are not as reliant on the outside temperatures. In the Arctic they are high for some, but the loss of ice creates new shipping routes and mineral opportunities too, as well as fishing and tourism.

The second answer junks all of this sophisticated analysis and sets the carbon price at an initial opening value, and then sees what happens. If there is a target, like net zero by a given date, once the price is arbitrarily initially set, it can be adjusted to rise and fall to whatever level is necessary to achieve that target. There is no *ex ante* attempt to get the 'right' price. Rather, the impacts of the initial price guide its revision. It is a bit like the setting of interest rates by central banks, only easier. Central banks have multiple objectives in practice; carbon is much simpler.

The great merit of the second approach is that it does not require detailed models and CBA, and it does not need armies of economists churning out econometric studies. In particular, it does away with all the sophisticated Integrated Assessment Models, striving for the perfect answer.[8] Indeed, it does not even require the marginal calculations which are in any event highly suspect for a system problem like climate change. All the work is done in setting the initial targets, and then the carbon price is just the tool to help to achieve them. The market *reveals* the costs.

In setting the initial opening carbon price, there are some factors that need to be taken into account. Since the purpose of the tax is to change behaviours, there needs to be time to adjust. Some of the capital stock, like cars, boilers, communications devices and machinery, turns over every few years. Other bits, like houses and buildings, last a lot longer. An immediately high price will yield lots of money (because people and companies cannot instantly change their capital equipment and just have to pay), but not much in the way of actual carbon reductions. The trick is to start low, but credibly signal that the price is going to go up as high as is necessary to achieve the target. In the case of net zero, this is a gradual ratcheting-up of the target and the expected carbon price over the next 30 years. The credibility of the commitment to 'do what it takes' means that the adaptation will ideally happen in advance, and hence the actual tax may not need to be so high. Indeed, if and when net zero is achieved, the price of carbon should in theory be close to zero.

The less credible, the higher the costs of the transition. In the absence of explicit institutional mechanisms, which we come back to later, the only comfort investors and households can otherwise have is the reliance of government on the revenues (provided they are not hypothecated). They need the money. This in part explains the endurance of the UK's carbon floor price, despite heavy lobbying by industry to weaken it.

The domain of the tax and the carbon border price

What exactly should the carbon tax be on? The answer to this depends on the targets and their specification. If the targets are intended to reduce global warming and, in particular, if the net zero 2050 target is intended to mean that the UK will no longer make any contribution to causing *global* warming, then it has to be on consumption rather than production.

This means that a carbon border tax should be added to the domestic emissions tax, and at precisely the same rate. It would then be legitimate to claim that, at net zero, the UK would no longer be causing further climate change. To start to unwind the damage already done would require a step further: to be net negative, in an effort to undo the largely coal-based carbon we have put in the atmosphere since the Industrial Revolution.

There are three objections to a carbon border tax.[9] The first is that it would lead to an increase in the prices of many imports, and this in turn would raise the cost of living. This is a strong political motive for trying to gloss over imported carbon and pretend it is all about domestic emissions. But the impact on the cost of our consumption is precisely the point: it is our excessive consumption which is doing the damage. Convenient though it may be for politicians to turn a blind eye to this source of emissions, it does not get us off the hook. If we do not make the polluters – us – pay for the emissions that are caused by our consumption of carbon-intensive imports, we merely delude ourselves.

Worse still, not to have a carbon border tax undermines not only the climate impacts, but also our domestic industries. Decarbonising to net zero domestically, but excluding imports, means that domestic production of everything from steel to beef faces a competitive disadvantage. It is one reason why both steel and agriculture have needed subsidies, exclusion from domestic carbon taxes and protectionist tariffs. Recall too that it is also one reason why, despite the fall in European domestic emissions, global emissions kept going up over the 30 wasted years.

The second objection is that carbon border taxes are impractical.

How, it is asked, can the carbon composition of an imported pack of bacon or a tonne of steel or fertiliser be measured? Recall how hard it is for you to fill in your carbon diary, and then imagine doing this for all imports. The answer is that, on a detailed tonne-by-tonne basis, emissions for each item cannot be practically quantified.

Fortunately, this does not much matter. It cannot be done domestically with precision either, and it does not need to be. Not to have a border tax, while at the same time having a domestic carbon tax, is *precisely wrong*. What we want to be is *roughly right*, to go in the right direction. Practically, the place to start is with the really big carbon imports. Apart from oil, gas and coal, which are relatively easy to measure and tax, the manufactured items that matter are a small number of ubiquitous inputs to domestic production. These include fertilisers, cement, steel, petrochemicals and aluminium (the 'big 5'). In addition to these, information and communications technology (ICT) is the coming big one, to power all the computing behind our digitalising economies (and Bitcoin too).

For these big 5, once the country of origin has been declared (as it has to be anyway), some simple metrics can be applied. For example, take steel from China. We know that coal represents around 60–70 per cent of its electricity generation, and we know roughly how much coal is used in blast furnaces. We could impose a schedule of carbon tariffs that reflect this. Steel from China could attract a tax on the assumption of, say, 80 per cent coal and a standard steel efficiency number. Each of the big 5 could have a 'going rate' by country of origin, and importers could apply for lower rates if they can prove that the carbon content is significantly lower. That is, after all, how tariffs generally work. The 40 per cent tariff on beef imports into the EU discussed in the context of a 'no deal' Brexit, for example, was not based on a precise measurement of the impacts on the domestic market, and there is nothing precise about Chinese, US and European tariffs on manufacturing goods. Approximation, not precision, is all that is required.

For very large industrial plants, the emissions in the country of origin should in any event be measured, as they have to be here. A UK steel mill had to buy carbon credits for the EU ETS, and now will have to buy UK ETS credits. The importer would fill in these details at the port. Compared with the normal customs requirements (and VAT administration), this is pretty trivial as an administrative expense, and proportionately very low, since cargoes of the big 5 tend to be high-value. Those polluters who import into the UK, and those importers who benefit from not paying for the pollution they cause, have an obvious incentive to protest about the complexities. They would, wouldn't they? But they protest too much, and are best ignored.

A final objection is the claim that border taxes would be illegal under WTO rules. There are two obvious responses to this. First, WTO rules do in fact allow for environmental adjustments, and so the border tax on carbon, tied to an explicit environmental target, and applied on the same basis domestically, would probably be fine. Second, the WTO faces far greater challenges to its authority anyway from the US/China trade wars. It is hard to imagine that, in its current dire state, it would want to legally challenge one of its members for taking action on carbon imports, and especially given the UN-led Paris Agreement which sets the overall global objectives, and the subsequent agenda for Glasgow. Not to tax carbon is to distort world trade. Carbon border taxes level the playing field, provided they are applied at the same rates to imports and domestic production. The WTO should be advocating carbon border taxes in the name of fair trading arrangements, not standing in their way.

The really great thing about carbon border taxes is that they also encourage a bottom-up incentive for other countries to introduce their own carbon taxes.[10] Think about the position of the country sending carbon-intensive imports to the UK. The tax is collected at the border and paid to the importing country's finance ministry. But what if the country sending the imports had its own carbon tax domestically, set at roughly the same level? Say the Chinese steel mill paid a carbon tax at home to the Chinese government. It would qualify for an exemption from the carbon border tax since the polluter is already paying. The incentive on the Chinese is simple: better the money goes to its own government than to the importing government. If enough countries opt for the border tax approach then there will be a critical mass and these powerful incentives will cascade through the global economies as every trading country now has an interest in putting a domestic carbon tax in place. A successful border tax is then one that withers away, a victim of its own success. It internationalises the carbon price.

This has a very important and radical consequence. It helps to solve the free-rider problem that Kyoto, Paris and Glasgow throw up. Instead of trying to persuade countries to sign up to top-down targets, this is a bottom-up approach and needs no grand international treaty. It is the only way to underpin national unilateral policy, by ensuring not only that the country with such a target stops contributing to global warming, but at the same time encourages others to take measures which will help them to do the same. This is about as good as it can get.

Upstream or downstream: where to levy the tax

The next question is where in the carbon supply chain to fix the tax. Should it be on upstream oil, gas and coal, or downstream on steel, fertilisers and plastics? Should it be on producers or consumers?

As with most taxes, the trick is to be pragmatic. Imported oil, gas and coal can pick up the carbon tax when they are used, which means that the tax falls on petrol and diesel at the filling station, and on the power station when it burns the coal and gas. Heating oil and gas can similarly be taxed on use. Imports could be taxed at the border. This is all very familiar territory. Primary energy is already taxed. The trouble is that it is done badly, and coal in particular tends to escape altogether. The taxes should be set on the carbon content of each. A tax on coal that reflects its carbon content, and with all the other pollution coal causes taxed too, would help to drive coal out of the energy mix. That in itself would make a big difference to climate change.

It still leaves taxes on imported products which used fossil fuels in their countries of origin. The border tax is what it says – a tax at the border. We already do quite a lot of this, notably on VAT. VAT is simplified by having a small number of rates. In the case of the carbon border tax, think particularly of these initially as different rates on each of the big 5 polluter products. They can be added to as experience is gained, getting gradually more roughly right.

Who fixes the price

Fixing and revising taxes is currently mainly a finance ministry function, but this is not necessarily the right approach for carbon taxes. It lacks credibility. The task of government is to fix the targets – such as net zero by 2050. In the UK it is the job of the CCC to come up with five-year rolling carbon budgets which break the overall target down into bite-sized chunks. Parliament has willed the end, democratically. Delivery is an agency problem. The CCC is one such body which could solve this problem.

It is for government to decide what the instruments of policy should be. It decides what and who to tax; and it decides whether the tax will be levied at the border. The delivery body takes the tax design and the carbon budgets as *given*, and adjusts the carbon tax on, say, an annual basis. It might also give guidance on what it expects the tax to be, say, five years out, projecting over the carbon budget set by the CCC. By giving clear guidance and an explanation for each tax change at the annual reviews, the market gets an element of stability in making its investments. The annual carbon tax review might be appended to the annual progress reports on the carbon budgets. The key thing is to make sure it is simple and credible.

The temptation for the Treasury will be to keep control, and indeed that is what is happening with the existing energy and carbon prices. When drawing up the terms of reference for the Helm Review in 2017, the Treasury insisted that I could not make specific recommendations about carbon and energy taxation. But what it could not do was evade the extra costs the absence of such taxation causes: the market will be uncertain; there will be lots of lobbying by vested interests; and the consequence will be that the carbon tax will have to be higher for any given carbon reductions. This is what happened when the Treasury set the interest rates, and it is why this function was delegated to the Bank of England. In the carbon case, the 'signal' of possible cuts in the Fuel Duty to appease motorists ahead of the 2019 general election is exactly the sort of short-term gesture that undermines credibility.[11] It may buy votes now, but it cannot avoid the consequences: increasing emissions and undermining its parallel plans to encourage motorists to switch to electric vehicles.[12] No wonder people keep buying SUVs powered by diesel and petrol. Treasury control of the means risks undermining the end (reduced carbon emissions) and increases the costs.

The key to this trick is *credibility*. Why should you believe that some future government will stick to the targets and be willing to allow the carbon tax to rise? Might not a future Trump or a Farage come along at some stage and simply ditch the tax in a populist gesture?

What to do with the money

Carbon taxes have the potential to raise a lot of money. It is even possible that they could raise almost as much as income taxes or VAT. Over time, the tax-take should fall as emissions fall. As with the border component, the tax should be a victim of its own success. But this happens later on: in the short term the emissions remain and hence the number of carbon units times the tax produces lots and lots of money.

What to do with it? There are two broad options. The classic public finance approach says that taxing and spending are two different activities, and that the Treasury should raise money as efficiently as possible through taxes, and quite separately decide how it is spent. The carbon tax revenues go into the general government pot, which is then spent on the various things that the government decides, such as health, education, defence, and so on. Among these spending items will be the public goods relevant to net zero, but each will be judged on its own merits independently of the level of the carbon revenue.

With this approach in mind, and if the government has a given total spending target, one use of the carbon tax revenues might be to lower other taxes. The government could rely on taxing bads, like carbon, rather than taxing good things like labour, and hence switch from income taxes to environment taxes.[13] This would improve the overall efficiency of the economy. In practice, however, governments struggle to persuade the electorate even to vote for proposals to raise enough tax to cover expenditure, and are instead forced to borrow to plug the spending gaps. Carbon taxes are likely to be used for additional revenue and not to displace other taxes. One implication of this is that carbon taxes may be introduced and raised not because the government of the day is convinced that they are needed, but because it has run out of other taxation options. Better the right outcome for the wrong reasons than the wrong outcome.

The alternative approach is to hypothecate the carbon tax revenue. Some argue that when it comes to climate change there is a get-out-of-jail-free card: we could tax carbon, but then hand the money back to all of us.[14] The relative price of carbon-intensive goods goes up, but our incomes are propped up by handing back the tax revenues. It is true that the relative prices will be right, although by maintaining overall consumption, they will probably have to be higher to meet a given target like 2°C. The problem comes when the redistribution details are addressed. Who gets the money? The poor, with the highest propensity to spend, and probably to spend on carbon-intensive goods? Or will it be a relief of income tax or VAT, and on which levels and classes? The carbon tax both reduces aggregate demand and changes the relative prices. The recycling just changes the relative prices, keeping up total spending. In the longer term, as there are more low-carbon goods and services to buy, this gets easier, but not in the short-to-medium term when the economy is overwhelmingly carbon-intensive. For many goods and services there are no obvious low-carbon alternatives.

An alternative hypothecation option is to spend it on carbon-reducing investments. Among these could be the infrastructure public goods and R&D. It could be through a 'carbon fund' or, if other environmental taxes were included, a broader 'nature fund'.[15] The impact of such a recycling approach would be enormous, and it would have the paradoxical consequence of the spending on lowering emissions leading to a lower initial carbon tax. Say, for example, the tax revenues subsidised renewables. A lot more would be built through the subsidies (which are a negative tax) paid out of the climate or nature fund. Emissions would be lower and hence the tax rate needed to hit the target could be correspondingly lower.

This is very appealing, and especially to those who benefit from the subsidies. But there is an obvious catch. The polluters who ultimately pay these taxes – you and me – would be worse off by the value of the tax, and then we would need to pay the usual taxes to get the spending on other things like health. It is a significant increase in the total tax burden, making us poorer than we otherwise would have been.

But then that is the point. The reason carbon should be taxed is that the polluters like you and me should pay for the costs we cause through our over-consumption. If we face up to our pollution and pay for it, we will be worse off. We will have to live within our environmental means. That requires sustainable consumption, not polluting consumption.

Raising the tax burden overall is what politicians try to avoid, for the obvious reason that voters don't want to pay. When the two approaches to the use of the revenues are compared, the important political issue is what the total tax burden should turn out to be and, in particular, how much the voters are willing to pay to address climate change. If governments were to decide that the UK should unilaterally stop contributing to global warming at some future date, that means net zero carbon consumption. The amount of money required to get there – the cost – is then determined by the route chosen. If an efficient path, driven by a carbon price, is chosen, this sets the minimum cost for the transition. Any other approach is more expensive and hence voters will pay even more in total to meet the overall net zero consumption target. Not to choose the carbon tax route with the border adjustment is to choose a higher-cost route, and not being willing to impose these costs means that the government is not serious about its overall target. In turn, since we vote for the government, we would as the ultimate polluters be signalling that we are not serious about climate change. We obviously should be.

6

NET ZERO INFRASTRUCTURES

All current infrastructures have evolved to serve and facilitate the carbon economy. They are no longer fit for purpose. A decarbonised economy will have to have decarbonised infrastructures to support it. The electricity grids will need to serve renewables and perhaps nuclear; the roads network will have to facilitate electric and perhaps hydrogen cars; the rail network will take the strain off aviation; the mobile and fibre networks will have to facilitate the smart electricity and transport networks; and the ports will have to serve low-carbon shipping. Doing all these is the second part of a unilateral strategy.

This is an infrastructure revolution in the making, to deliver mid-twenty-first-century networks. Nothing like this has been attempted in one go over just 30 years. Only wars provide anything like as dramatic a challenge, and almost always within the context of given technologies. To give some scale to the investment required in the UK, HS2's budget has been marked up to over £100 billion. It is not hard to get to £1 trillion, and possibly a lot more, for all of the above.

Markets are not enough

These infrastructures will not be provided by private markets on their own, and to the level and extent that meets the public needs for the economy as a whole. They will need state intervention, and much of the climate change debates over the last 30 wasted years have neglected these broader infrastructure issues. Private savings and tax revenues will need to be diverted to these nationally planned necessities. Only the State can drive a transformation on this scale. To pretend otherwise is to not take climate change seriously.

The reasons private markets will fail to provide enough of the right sorts of infrastructures are multiple: they have important *public goods* dimensions; there is the USO; and there is the exposure of investors to possible expropriation through regulation and nationalisation.

Recall that a public good has two core characteristics: it is non-rival and it is non-excludable. The non rivalry reflects the fact that the marginal cost of each extra user is zero up to the point the networks are congested. This means that, provided the networks are invested to a level sufficient to meet demand, the best outcome is one where everyone gets to use them, so that every bit of extra benefit is grasped. Since the networks are also essential services, and any decent society provides them universally, any economy that does not suffers lots of inefficiency as a consequence. This is why governments always have to intervene to protect poorer, vulnerable and more rural customers, and why they are all intervening to complete mobile, broadband and fibre networks. This is the USO.

The problem for infrastructure markets is obvious. In theory, the efficient answer is to set the price equal to the marginal costs, but no private investor is going to provide sufficient infrastructure capacity faced with marginal cost pricing. How are they going to recover the cost of investing in building the networks in the first place? It is a classic infrastructure problem: the average costs (representing the fixed and sunk costs) are high; the marginal costs low.

Any private sector provider would need to recover these fixed and sunk costs. But how, without deterring customers? The answer has to be state intervention, guaranteeing that private investors get paid or by transferring the burden of these fixed and sunk costs onto taxpayers. This is where regulation comes in. Even if the State promises that investors will get paid back, what happens when the investment has been made, and the costs sunk? Could the government renege on its promise, and screw the prices down to the zero marginal costs?

Investors obviously need some protection and, post-privatisation, this takes the form of a contract with the State to the effect that their assets will not be expropriated. It is, in effect, a guarantee that they can earn a reasonable return on their regulated asset base (RAB), and this is a way of recovering their fixed and sunk costs. Customers are made to pay the average, and not merely the marginal, costs.

How can this be achieved? The reason customers pay is that they are forced to, and that is because the networks are typically monopolies, natural monopolies. It is not efficient to pay to have two networks because of the enormous economies of scale. When these costs are sufficiently high, some may be tempted not to use the services at all. Opting out of the electricity network is a real possibility now, something that governments are very keen to head off, since it reduces the number of people and companies making their contribution to the fixed costs. So paying for infrastructure networks has to be compulsory, either through user charges which recover the operating and RAB costs, or through taxation.

Regulators and politicians are always vulnerable to popular opinion, and the public always wants cheaper travel, broadband, electricity, water supplies and sewerage services. Investors know this. This becomes critical when there is fast technological change. Take the communications sector, critical to a smart decarbonised world. BT owns lots of legacy assets, and these sunk and fixed assets have been charged to pay BT's dividends and to fund its pension deficit.[1] If you've ever wondered why Britain now lags behind many other developed countries on broadband and fibre, this incentive problem is one crucial reason, and why BT wants to be guaranteed a long-term return on fibre for 20 years. It has finally won this argument.

Taking the fundamental public goods problems, and the legacy assets incentives such as those in the legacy network example above, it follows that a modern efficient economy will have to have the extent of its networks set by the State. Government is responsible for defining how much spare capacity these networks should have and, as a result, for putting in place mechanisms to achieve these capacity margins. Government also has to ensure that investors are not expropriated, and must find ways of doing this that do not frustrate investment in new infrastructure and new technologies. Without these interventions, the economy will be stuck with substandard assets, investors will focus on sweating them rather than investing, and technological innovations will be slow in coming to fruition. Next time you try to make a mobile call along even some of the major rail routes in Britain, wonder why sewage spills into the rivers when there is a storm, and stand on a railway platform waiting for a train to run along the mostly empty railway lines, this is the reason. It is government failure and poor regulation of private utilities.

The centrality of infrastructure networks

Imagine what a net zero economy would look like. There would be lots of decentralised renewables generation, possibly some nuclear power stations (both large and small), smart meters, smart devices, interconnected homes and the internet-of-things, autonomous electric cars and perhaps hydrogen-powered vehicles and electric trains. Travel, especially by air, would be much reduced, and holidays would be much more local, as would quite a lot of food production. There would probably be more remote working, including from home using video links, as many people had to do during the Covid-19 lockdowns.

Confronted with transitioning to this low-carbon world, now think about the existing network infrastructures. The electricity system is designed around ever-larger power stations (coal, nuclear and now gas) transmitting electricity to the local distribution networks and then your home. As yet, smart meters are not fully in place (and some do not fully work); there is no clear understanding of how to use (and who can use) the data; and smart appliances are a long way off becoming universal. The reason this smart technology is not in place is because the communications infrastructure is not up to the job, and nor will it be for the whole country for perhaps another decade. You cannot run a smart meter or enable your smart devices unless you have good internet and mobile connectivity.

The road system is designed entirely around petrol and diesel vehicles. It is anything but smart, and incapable of supporting the roll-out of smart cars and autonomous vehicles. Charging points for electric vehicles are still notable by their absence even in major conurbations. Where they are available, the roads are often so congested that getting to a charge point can be a challenge in itself. The oil companies have not developed a retail petrol and diesel network designed around the electricity grid, for the very good reason that it has been irrelevant. Much of the railway network still relies on diesel, and when there are power cuts the electric trains grind to a halt. Airports continue to be expanded, and many if not most people get to airports by conventional cars and buses.

The electric car example illustrates why this is a chicken-and-egg problem. Motorists are not going to buy an electric car if they cannot be confident that they can readily charge it. Investors in a private electricity distribution company are not going to invest in advance of demand, for fear that the revenues will not be there, and in any event the new technology may yield very uncertain outcomes. Building an electricity charging network requires government intervention, to pre-invest ahead of demand with regulatory guarantees to ensure that the costs and risks are properly paid for.

Why does this not happen? Why don't we just get on and get all this stuff done? The reason is that pre-investment means paying now for future potential benefits. It means that the capital expenditure of the companies goes up, and therefore so does your bill. But voters say they want cheap electricity that is secure and low-carbon, not more expensive electricity to provide sufficient capacity for a future without the carbon. It is just another example of polluters (you and me) not being willing to pay to clean up the mess we are creating. We cannot have a decarbonised economy without the supporting green networks, and we cannot convert from fossil fuel networks to low-carbon ones unless the investment is made. We cannot invest without savings and a credible guarantee that customers and taxpayers will actually pay up.

We could have an electricity charging system like Norway and its associated smart meters;[2] we could have a high-speed electric railway system like France; and of course we could have fast fibre like Spain. This would have one other advantage: it would be no regrets. We need all of this anyway.

A net zero national infrastructure plan

A net zero plan starts with these core infrastructures as the backbone of the low-carbon economy. It starts with an overall network infrastructure *plan* and then works backwards to the necessary investments. The first priority is fibre and the communications networks. These infrastructures, new in the last 30 years, enable everything else. They have to be able to handle vast amounts of data, and they have to be secure and safe from cyber-attacks. They have to be over-built, with lots of deliberate redundancy in them. Building just to meet demand as and when it materialises is not enough.

Governments face mixed incentives here. On the one hand, there is a clear recognition that these are new universal services and, like electricity and water, citizens are entitled to them. There is an emerging USO.[3] It is increasingly difficult to participate in society without a good internet connection, and it is close to impossible to be economically efficient without access to internet banking. Almost everything is going online, in a world where not everyone is online. Citizens need these USOs to be part of the modern decarbonising world.

Instead of promoting a very late-twentieth-century idea that national infrastructures are best defined by competing companies and competing networks, what is needed is a straightforward plan for fibre. The plan should set out what is required for the new systems that will rely on it, and the USO. The data requirements for electricity, roads, railways and so on can be estimated, and the systems built to service these. Since the risks of having too little capacity are much greater in their cost implications compared with having too much, a healthy over-capacity should be built into the plan. What is then needed is a strategy to implement the planned requirements, and someone in charge of delivering it. The practical necessities are largely absent from the National Infrastructure Strategy.[4]

A credible national plan should fill in the details of what this communications infrastructure will be serving. Lots of people have thought about this already. It is not hard to describe the electricity system that is evolving, and we will do more of this later on. It will have a lot of decentralised renewables generation, an active demand side and lots of storage. It will serve not just direct electricity consumption, but also electric vehicle batteries and household heating. A proper national infrastructure plan has to set out the investments to turn the electricity networks on their heads, starting at the bottom – the very local – and working upwards. In this it also has to incorporate the new challenges of intermittency and the interactions between the local networks, storage, local generation and batteries. The government's Ten Point Plan and Energy White Paper only scratch the surface of what is required.[5]

Transport is typically considered a separate issue, and this is reflected in the companies and in the structure of governments. Transport ministers tend to focus on spending on potholes, road improvements and handing over ever-increasing sums to the railways. There have been attempts to plan transport as part of the wider climate change objectives at a time when electric vehicles were only a distant prospect.[6] The climate strategy then was to get from roads to rail, not to decarbonise roads and hence perhaps even encourage switching in the opposite direction. Fast forward to the present and it is not at all clear that smart, autonomous electric cars will be worse than trains, and the former have a lot more flexibility than the latter. It might well be that trains become more narrowly focused on very local city transport, and long-distance high-speed trains substitute away from air travel and airports. That, however, has not stopped the Department for Transport rightly arguing that there must be a big shift towards public transport in its plan for decarbonising transport.[7]

The concept of a plan was right then and it is now: it was just that past transport plans have typically come without the necessary funding and without being joined up with the rest of the energy systems or with communications. Transport is increasingly becoming electricity plus smart communications, rather than diesel plus hand signals.

Rather than attempt to predetermine the transport mix between the different modes, a new plan would focus on making both road and rail networks fit for a decarbonised world. The result may be too many railway lines and even too many roads, but in the broader scheme of the economy as a whole the extra costs would be lost in the economic noise, and in practice both sets of capacity would probably get used.

The most bizarre failures of UK national infrastructure planning come with the decisions on HS2 and on Heathrow and other airport expansions. HS2 is eye-wateringly expensive – now estimated at £108 billion, or even more in current prices. You can buy a lot of infrastructure for this sum, and remember that in the end we will have to pay for this through user charges and taxes, and hence have less to spend on other things, including other infrastructures. Full fibre, by contrast, might cost around £30 billion. There is little doubt that HS2's appeal to politicians has over time been one that might be called the 'trophy project syndrome'. Having decided to support it, there have been repeated attempts to find rationales for it. The first was 'time saved', and to set the value of time saved by the higher speeds against the costs. If it is assumed that all time on trains is wasted, then it is simply a matter of taking the number of minutes saved and multiplying by the number of passengers and the relevant wage rate. This produces a positive economic value for the project, but not much. Next to be claimed are the economic growth benefits. High-speed railways are supposed to produce 'growth nodes', and the extra GDP is then set against the cost (and anyway the costs are actually extra GDP as the spending reverberates around the economy through the multiplier). Then there is the rebalancing of the North against the South, although quite how making it even quicker to get to London is going to reverse the trend of recent decades has never been explained. Each review of HS2 tries to come up with more novel justifications.

The best rationale for HS2 is that it would be part of a high-speed system connecting Edinburgh and Glasgow via Manchester and London to Paris, Frankfurt and Milan. In this European high-speed rail system, the benefits from HS2 would accrue across Europe. This system could then cut into the European regional airline market, and passengers would be able to go by train rather than plane to a host of destinations in Europe. It could deliver very important carbon benefits as part of the strategy to reduce aviation and its pollution.[8]

Except that HS2 is not now going to be connected to the European system. It is going to stop at Euston, leaving passengers to walk or take the tube to Kings Cross to join the Eurostar. Or it might even end up to the east of London at Oaks Common. Why? Because the government at the time decided that it wanted to cut the costs, and could lop off £500 million from the then £50 billion budget by stopping at Euston. Once built, it will barely reduce emissions in Europe over the rest of the century. It would be hard to make this up.

Perhaps even worse is the approach to airport capacity and to Heathrow's expansion. The argument for a new runway is that the demand for air travel is outstripping runway capacity, which (pandemic aside) is true.[9] Hence it is concluded that supply should be expanded to meet the demand. For a government committed to net zero this is an extraordinary approach. The starting point should be the demand for aviation *consistent with the net zero target*. This should be on a net carbon consumption basis: Heathrow's expansion would lead to more emissions for the global airport network, and not just on the ground or over UK airspace. (It is therefore perhaps not surprising that, when in February 2020 the Court of Appeal ruled the proposed expansion illegal on the grounds that it is inconsistent with the government's own commitments to tackling climate change, it was decided that the verdict would not be appealed.) Take a trip to Heathrow on a normal day and just look around you. It is clearly at the limits of its capacity. Can we have this level of air travel and decarbonise to net zero? Clearly not. Would demand be so high with a proper carbon price? No.

A net zero strategy will require the aviation sector to contract. It is not enough to reduce the ground emissions by putting more public transport to and from airports in place, although a glance down immediately after take-off from Heathrow at the sheer scale of the car parks (a major revenue source for many airports) indicates that this will be important until all cars are electric. Noise matters too of course, as do the air quality issues over and above those of carbon emissions. Like burning coal, aviation causes multiple pollutants. It is very dirty. While it might be argued that aircraft will eventually be low-carbon, perhaps fuelled by hydrogen, genuinely low-carbon biofuels or by batteries, few can imagine this happening at scale anytime soon, and certainly not before the 2050 net zero target date. Aviation and net zero just don't mix well.

Sorting out transport, electricity networks and, most importantly, communications is a necessary condition. No minister can stand up and say they have a credible net zero strategy without this. The economic costs of this bit of decarbonisation are somewhat ameliorated by lots of extra non-carbon benefits. Imagine what you could do with full fibre in your home. Imagine all the businesses that could move out of crowded cities so that their employees need not commute on crowded trains. Imagine the air quality improvements that might follow from all these transport changes. The Covid-19 lockdowns have given us a glimpse of some of these.

This infrastructure plan would need to take account of new demands, some of which will come as a result of climate change. Two spring to mind: desalination to provide fresh water; and air conditioning to tackle the heat. Both are electric technologies.

Major cities around the world are coming to rely on desalination. The trouble is that it is massively energy-intensive. The challenge is to provide enough renewable generation of that electricity, and to technologically improve the osmotic processes to separate out the salt. Given that the desalination units do not have to be 'always on', it is possible that they could absorb surplus renewable energy, from both wind and solar. Imagine a world in which water becomes much more widely available to all those cities on the coastlines, and especially those cities in places like the Middle East with its intense sunshine.

In a hotter world, lots of emphasis will be placed on keeping cool. Air conditioning is an electric technology that lends itself to being used in the daytime and does not need to be 'always on'. Refrigeration has some flexibility too. It can also use surplus wind and solar energy. Adding desalination and air conditioning to the new electricity demands from a digitalising world makes for an increasing demand for electricity going forward. While buildings and machines may (and will) become more energy-efficient, the demand for energy is nevertheless likely to rise. Indeed, the more efficient its use, the lower the cost of each unit, and hence the *higher* the demand for any given household or company budget.[10]

These new demands reinforce the point about building excess capacity margins into a national infrastructure plan. We are more likely to be surprised on the upside by electricity demand than find that energy efficiency reduces demand. Better to put the capacity in place.

A regional and global infrastructure plan

Climate change infrastructure is not just national: it is regional and even global. We have long been familiar with the international oil and gas networks, and the coal terminals. Liquefied natural gas (LNG) has added a whole new infrastructure for LNG supertankers and terminals. Less obvious is the importance of electricity networks, and this has partly been driven by the limitations of the transmission technologies. The superconductivity once heralded as bringing a bright new future has not emerged as promised.

Notwithstanding the technical limitations, there has been a gradual growth in the interconnectivity of electricity networks on a regional basis. Most European countries are connected to their neighbours, with France's 80 per cent nuclear capacity as a pivotal European supply, especially to anti-nuclear Germany. The Nordic countries have brought hydro to balance the intermittent wind in Denmark and northern Germany. Looking ahead to decarbonisation, the scope and scale of this international infrastructure is potentially a big deal. Consider, for example, the abundant geothermal heat in Iceland and its capacity to generate electricity. It could supply the UK, and from there other parts of Europe. Wind generation in northern Norway and throughout the North Sea can supply intermittent power to Europe, as indeed Denmark currently facilitates.

As an alternative to electricity transmission, Iceland (geothermal) and Norway (wind) could provide the feedstock for a hydrogen economy through the process of electrolysis. In Iceland, there is potentially almost no limit to the scale with which this could be done. If hydrogen takes off as a viable fuel, initially for the harder-to-reach transport of ships and larger vehicles, and perhaps also trains (and even domestic heating), then its manufacture will, like oil and gas, be concentrated in specific geographical locations. If the hydrogen can be transported readily (possibly as ammonia), it can be made in remote locations, using not only geothermal but also large-scale hydro power. All of this will require new infrastructures, replicating what has happened to turn local gas production into a global business.

The potential for geothermal and wind, both directly as electricity and indirectly as hydrogen, could be considerable in particular regions. Solar offers a global supply of electricity and heating on an altogether bigger potential scale. Think of the cloudless skies over the Sahara. Think of the Gulf states, like Saudi Arabia and Iran. Think of central Asia, and the southern states of the US. The supply of solar energy to the planet is for all practical purposes best regarded as infinite. The question is not whether there is enough, but how to get it from more remote locations to high-density populations in a useable form by developing high-voltage international transmission networks. Spain already functions on a limited scale in this regard. Imagine a high-voltage electricity grid across the top of North Africa, feeding into Spain, Italy and the Balkan countries, and these countries in turn pushing electricity into the core of Europe, to France, Germany and the Netherlands.[11]

While much ink has been spilt on the renewable technologies themselves, they will be of limited value without access to the markets that demand the energy. This is what the IEM in Europe has been trying to do by interconnecting between EU member states. It is a necessary step to integrating with networks outside the EU. It requires initiatives to make sure that the infrastructures are in place, and to share the costs. The reluctance of some major European countries to allow the EU to have some control over their systems, and instead pursue a narrow nationalism, is a significant obstacle to winning the wider decarbonisation prize. Brexit makes it very difficult for the UK.

System operators and implementing the plans

Infrastructure comes in systems. It is not an atomised collection of individual investments, each considered in isolation from the rest. It is therefore not amenable to narrow CBA. Each wind farm has an impact on the security of supply of the system *as a whole*, and each connection to a wind farm opens up opportunities for others to join the networks. HS2 is a system, with atomised CBA having little relevance, and all the commuter networks work only if they are interconnected systems. It sounds obvious, but it is often forgotten.

Because infrastructure comes in systems and because, as identified above, there are never going to be the necessary private incentives to provide enough of it, and in advance of the things that are going to need to be connected to it in order to decarbonise on the net zero timetable, it has to be planned. There is no credible and efficient *laissez-faire* solution.

Planning a network is a constant process. New connections, new bits of network, and new ways to integrate data arise all the time, and there needs to be a continuous evolution of the networks to exploit the opportunities as they come along. A system needs to be coordinated, and that coordination requires a system operator.

It was once straightforward to solve this system planning and operation function. Most networks were state-owned natural monopolies. Throughout Europe, great companies like EDF, RWE, the CEGB, British Gas, British Telecoms and British Rail fulfilled this function.[12] In the US, monopolies in the private sector were regulated on a rate-of-return basis and commissions oversaw the systems and their developments.

In the 1980s, and especially the 1990s, all this changed as the new, more market-orientated and competitive utility models were promulgated. The great public utilities were privatised, and competition was introduced to serve customers who could (and were encouraged to) switch suppliers. The infrastructure networks became the carriers of the producers' supplies to the consumers, thus taking a passive role. No one planned the systems to any great extent anymore. They were whatever emerged from the myriad decisions of increasing numbers of competing private producers and suppliers.

In some cases, the remnant monopolies survived. Almost everywhere monopoly electricity network transmission and distribution companies remain. It was in railways and communications that the great disaggregation took place. But even where the networks survived, they were now supposed to passively do what the producers and suppliers wanted. National Grid, for example, no longer had much say in where power stations on the system were located, and little say in how the North Sea offshore transmission networks developed to collect the wind power and bring it ashore. For a while, neither did it have any say in how much capacity was needed. The market, it was believed, would sort all this out. The meters were taken away from the distributors (ushering in the great debacle that followed in making suppliers the lead for smart meter installation). The suppliers were split out from the distributors, and the networks unbundled from both generation and supply in electricity.

The results were predictable. The system dimension did not vanish just because it was largely ignored, and over time it became increasingly apparent that the competitive and liberalised markets could not solve all these system problems. Power stations were built in the wrong places; the offshore network became a series of bilateral links; and there was not enough investment to meet expected demand.

By the second decade of this century, the UK government had become the central purchaser of almost all new generation on the system (and much of the existing generation too), through subsidy contracting to renewables and nuclear, and through the capacity market. The liberalised generation market was effectively killed off.

To make this work, the system operator had to be reconstructed, and it is notable that almost all the major infrastructures relevant to the net zero objective now have explicit or implicit system operators. They need to, and these need to have plans, and these plans need to be consistent with the net zero targets.[13] At present this is at best work-in-progress.

As more system thinking and system planning returns, some argue that the great utility infrastructure networks should be brought back under public ownership and control. While this may be superficially appealing, it is neither necessary nor particularly desirable. What has been learned from 30 years of privatisation is that there are some things that competitive markets do well, and some things that are best left to public bodies. Governments are not very good at producing things. They are not very good at efficient road building or power station construction. These are best done by competitive, private, profit-driven companies, bidding for contracts. What governments are better at is the *coordination* of systems and their *planning*. Governments can take a broad economy view, and are not tied to the interests of each infrastructure network in isolation. They can comprehend the need to build fibre for decarbonisation, and hence for the transport and energy infrastructures.

For these reasons the various infrastructure core system operators should be public, and not private, entities. They are the public goods providers and it is inappropriate for private investors to decide what that public interest is. To leave these functions with the likes of National Grid or BT has the potential to create serious conflicts of interest. The Helm Review sets out how to do all this for electricity at the grid and regional levels.[14]

System planning and coordination functions matter most when there is rapid technological change. It is highly unlikely that the oligopoly of broadband and fibre companies will define and implement the optimal national networks to a timetable that meets the net zero objective. Indeed, they cannot even plan the mobile system to avoid the numerous blank or so-called 'non-spots'. What they can do is lay the cables and do the works. To imagine that a country will end up with the communications networks it needs for net zero without a plan or a coordinator is not just naive. It is a category mistake.

Similarly, consider the speed of technical change going on in electricity systems. New technologies are, as noted, turning the old centralised system with bigger and bigger power stations on its head. But who is going to ensure that the regional and local electricity systems are put together in a coordinated fashion, so that active and smart demand management, storage and batteries, electric car charging and networks are developed fast enough to facilitate the energy transformation that net zero requires? The current answer, which is to leave it to the existing distribution companies, is a big mistake. They have an enormous conflict of interest and will want to protect their investments in the grids and their associated RABs, analogous to BT and its legacy assets. It is a recipe for delay, obfuscation and the inhibition not just of the entry of new technologies, but of the wider electric car charging networks too.

Making it happen

There can be no net zero without appropriate infrastructures, and these are never going to be built if it is left to privatised companies alone. The private sector will deliver what is profitable, and what is profitable is partial rather than complete networks and systems. They will not complete the job, because they risk not getting paid and even being expropriated by regulators.

Governments need to plan the networks, and the new challenge is to plan these with net zero in mind. There are only 30 years left to do this, and on current progress it is not going to happen. There won't be enough capacity and it won't be on time. The last 30 wasted years have been characterised not only by the golden age of fossil fuels, but a dark age for infrastructures. It is not that we do not know what is needed. What we lack are the plans and the determination to make them happen. This requires us to be willing and able to pay for them, and provide the savings to underpin the investments.

7

NATURAL SEQUESTRATION, OFFSETTING, AND CARBON CAPTURE AND STORAGE

More carbon is going to be put into the atmosphere for decades to come, and probably for ever. To think otherwise is to live in a parallel universe. Net zero does not mean stopping all emissions: it means taking at least as much carbon out of the atmosphere as we are putting in, and from 2050. If you cause pollution by emitting carbon, through production or consumption, you should pay, and that payment should go to compensate for those polluting emissions. This is how you are going to turn your carbon diary into your own net zero consumption. It is the third part of a unilateral strategy, along with the carbon price and building the infrastructure.

The net gain principle goes a bit further: it says you should put back a bit *more* carbon than you emit, on the precautionary principle. Why precautionary? Because there will always be some leakage, and there will also be cheating. Measuring carbon imports as part of net zero carbon consumption is always going to be an approximate effort (roughly right rather than precisely wrong), and trying to capture everything from a garden bonfire to a plastic wrapper is all but impossible. Our net carbon consumption emissions are an estimate, not a precise measure. A margin for over-emitting, be it accidental or deliberate, is required in order for us to be sure that we really are net zero. Your carbon diary should be net negative, and not limited to net zero. There should be net carbon gain.

Why offsetting is required

Compensating for current emissions is a straightforward moral position to take. You cause the emissions, so you pay, with a precautionary bit more to make sure you really do. But what if it makes no difference to climate change? What if others take a different moral position? With a fixed overall *global* net zero target, the more you offset your emissions, the more room for others to increase theirs. This is, as already noted, what happened in Germany under the *Energiewende*. By investing in renewables, overall emissions were reduced. Yet a fixed reduction target left room to expand the coal-burn without breaching the limits. The more the price of the EU ETS permits fell, the more successful the renewables, leaving more room under a fixed cap for coal.

We will come back to this point in considering how net gain markets could operate. The moral force of the net gain principle can take this consequentialist line: that all that matters is whether the consequences are lower net emissions. But recall that a unilateral net carbon emissions target is not necessarily consequentialist. While it may be argued that the reason we should be net zero carbon consumers is that this will demonstrate to the world our commitment and provide an example of how to do it, and therefore have the consequence of encouraging others to do the same, we might also argue that we have a *duty* to do this, even if others do not. In other words, there are reasons for taking a net zero approach irrespective of the consequences. We have a duty not to damage the planet, or at least to tread lightly on it, so we are not responsible for further emissions, regardless of whether others follow.

How far back should we go? Is it just today's emissions that need to be compensated for? What about yesterday's? What about all the emissions since the Industrial Revolution? From the perspective of the Brazilians cutting down the Amazon rainforest, it is a bit rich to be told not to do so by Europeans who long ago cut their forests down, and also burned so much coal, and caused the largest share of the stock of carbon still in the atmosphere. As the world has moved from 275 ppm to over 400 ppm, how much of this is the responsibility of UK people in the past?

This is far from a purely academic question of documenting past emissions. High emissions in the past allowed the UK to pull ahead, and the Industrial Revolution facilitated the twentieth-century carbon economy. UK citizens today are much richer, on the back of those past emissions. It is analogous to the slavery argument, although at least the carbon polluters of the past did not know they were causing climate change, whereas the slave owners knew all too well the suffering they were causing. The question is whether the current generation should not only apologise for what happened in the past, but also further compensate the affected countries and even individuals. This is particularly pertinent if the countries that are going to be most adversely impacted by climate change have a significantly lower consumption level, including carbon consumption.

This 'sins of the past' moral argument has critical implications. Should we contribute to global funds to help others decarbonise and take adaptation measures? Do we have an obligation to enable climate refugees to live in the UK? All of this is carbon compensation way beyond the net zero line.

Types of sequestration

The past moral obligations are a matter of obvious dispute, because moral standpoints always are. What is not in dispute is that if the net zero carbon consumption target is to be achieved, there will have to be offsetting measures, putting back some carbon to compensate for what continues to be emitted. On this front, we haven't even started yet.

Natural sequestration (and natural emissions) happens all the time. Our interest is in *net* sequestration, over and above the natural emissions from anything that breathes out carbon, and from the erosion of rocks. Again, it is more demanding than it seems: any compensating sequestration for net gain must be over and above the natural sequestration that would have happened otherwise. It is no good cutting down trees, burning them and then claiming that replacing them is compensation. (We return to this point when we come to biomass, and in particular wood burning.)

There are two types of sequestration that might qualify for net gain. The first is through natural processes – *natural sequestration* – and the second through carbon capture and storage (CCS) – *industrial sequestration*. Natural sequestration includes planting *additional* trees, encouraging the recarbonisation of the soils and the regeneration of peat bogs.[1] It also includes the creation of new coastal marshes. CCS involves siphoning off carbon from power stations and industrial sites and then piping it into storage wells, such as old oil and gas fields, salt cavities, and possibly even into the deep sea. It can be in gas form or converted to liquids. Eventually it could even be directly from the air.[2]

In all cases, the baseline needs to be defined. What is the baseline forestation against which *additional* trees are planted? What is the soil baseline? Is it the soils after intensive chemical-based agriculture has stripped out the carbon, or is it the carbon levels in the soils that there would have been had these industrial farming techniques not been applied? When we come to considering agriculture, there is a great difference between paying farmers on carbon-poor, intensive cereal production land to put carbon back that they have allowed to be depleted without paying the carbon price, and rewarding farmers who, through their more sustainable practices, have already protected their carbon reserves.

For CCS, consider an oil company that is depleting an oil well. When the well starts producing, the pressure is typically very high. In some cases, it is so high as to risk blowouts, such as the BP Deepwater Horizon disaster in the Gulf of Mexico in 2010. Gradually the pressure drops, and when around half or even less of the oil has been extracted, it is typically so low that the well may be abandoned. Now along comes CCS, and the possibility of getting paid to inject carbon back into this well. Gas is a good way of increasing the pressure again, and hence the oil company could get paid to use the well for CCS, and get the bonus of more oil extracted. The former is a step towards net zero; the latter is not.

This might work out if the carbon tax is set at the right level and applied to the oil extracted. But it would not be a good result if there is no proper carbon price. The key point is that each of our three principles – polluter pays, public infrastructure and public goods, and net gain – have to be applied simultaneously to pursue net zero efficiently, and not in isolation.

CCS raises some further challenges as a means of offsetting carbon emissions. There are practical questions about volume and the number of possible stores. In the North Sea there are conveniently quite a lot of old oil and gas wells, now significantly depleted, and there is a ready-made pipeline system in place. The North Sea is shallow and relatively easy to work in. All that is required is to find a carbon source, reverse the gas flows into the wells, and then seal them up.

There are old coal- and gas-fired power stations and industrial plants on Britain's east coast. There is the Drax power station with its biomass boilers. The investment is needed in separating out the carbon pre- and post-combustion, and then joining up to the existing pipelines.[3] The wells themselves are gas-tight (which is why they could be exploited in the first instance), and sealing is a matter of plugging the drill holes that were created in the past. If there is a bit of leakage, as long as it is monitored, it is still a better option than simply venting the carbon to the atmosphere from the power station or large industrial complex.

For this reason, if there is anywhere in the world CCS might work, it is in the North Sea, and that is why, for global reasons, it should be tried. It is not, however, going to make a big difference to global warming; nor is it a get-out-of-jail-free card for oil and gas producers, let alone coal producers. The reason is that the volume of the carbon put back as CO_2 is typically greater than the volume of carbon extracted, so many more holes are needed than existing wells, and there may not be enough other holes readily available to utilise.

If CCS is to make a significant difference, the volume effect needs to be solved, possibly through creating concentrated liquids, and some other very large-scale storage needs to be developed.[4] There is discussion about using deep-sea trenches, with the pressure of the water at depth storing the carbon.[5] Yet it is far from clear how the carbon would get to these deep locations without leakage, and what its impact would be on the oceans, including the biodiversity and the acidification issues. If these sorts of CCS are required, they will take time to develop, be very costly and carry significant risks. In comparison, natural sequestration is readily available now, has low costs, and is very low-risk, provided that, for example, the right sort of trees are planted to take account of other environmental considerations like biodiversity and water management.

Net zero for companies

Let's go back to the challenges that face an oil company in tackling net zero, discussed in chapter 1, as the demand for oil and gas just keeps going up again after the Covid-19 lockdowns, and consider where sequestration might fit in. It might be BP, Shell, Total,

ExxonMobil or Equinor. Suppose it sets itself the objective of being net zero by a certain date (as most now have), and proposes to achieve this through global tree planting to offset its carbon emissions from all the oil and gas it produces. It might be an interim step as the company switches from oil and gas towards becoming a renewables electricity company, or simply closing down.

The company could, and probably should, do some CCS. But the scale of its emissions is so great that CCS is not going to do the job. It therefore needs to turn to natural sequestration, like tree planting. In theory, it is simple: each tree sequestrates so many tonnes of carbon over a number of years. Each tree costs a given (very small) amount. Each tree may – if it is the right sort of tree – yield other benefits to biodiversity, soil, water and flood protection, and so on, depending on its precise location. The oil company can work out the level of its emissions, then calculate the number of trees it needs to plant, and finally derive the total cost. Other polluters, like airlines and airports, can do the same. At first glance, it looks practical and cheap.

There are some obvious issues. The trees take time to grow, and in the early years there is a process of waiting. The polluter will be emitting carbon faster than its trees are sequestrating, so it will be building up a carbon debt and will require more trees in the future to compensate. Some of the trees might die, and so a margin for this needs to be included and, more broadly, the company cannot be certain that the host country will not later cut them down for development and agriculture. The additional precautionary element will have to be significant.

So far so good. But recall our earlier Brazilian example, and let's return to the baseline. These trees have to be *additional*. Imagine the trees are to be planted in Brazil, augmenting the Amazon rainforest and making it bigger than it otherwise would have been. This is a tricky counterfactual because the Brazilians are accelerating the cutting-down of the forest. As fast as the polluting company might be offsetting its emissions by planting trees, Brazil is speeding them up.

Would it count if the company chose an area of forest that Brazil had cleared to raise cattle? This would be like Canute on the beach trying to stop the tide. Worse, as noted earlier, it would provide an additional incentive to cut down the Amazon to get paid to plant it back again. This would not be an appropriate baseline.

Our polluting company might therefore turn to planting trees in Europe. There are large land areas with no trees, so surely this would be a net gain? Yet the north European plain was once covered in trees, as was most of the UK, including the great Caledonian pine and birch forests of Scotland. The only difference between Brazilian and European tree planting is when the trees were cut down.

If our polluter just paid the Brazilian government not to cut the trees down it would probably have a bigger effect on emissions and sequestration than if it planted new trees almost anywhere in the world. Yet it would not be an example of natural sequestration of its own emissions as part of its efforts to offset its pollution. It would be a global payment to stop the emissions from Brazil that would otherwise add to global warming as the clearing continues. It is an emissions mitigation strategy, not a sequestration one.

Thinking about it this way, it is not clear that there is a sequestration option for our polluter by planting trees. The existing trees should be preserved anyway, regardless of its emissions. New trees should also be planted anyway, regardless of the company's emissions. Narrowing the options this way indicates that the company's scope for carrying on pollution by offsetting is pretty small. The trees are needed in order to get to net zero, and the companies trying to get to net zero should be cutting emissions. The carbon tax – the polluter price – is what should do the work here.

But you might be thinking, is it not a good idea to have polluting companies like the major oil producers paying for the tree planting, perhaps somewhere not as controversial as the Amazon? Are not financial flows like this exactly what is needed to get the trees planted? If, for example, the new Great Northern Forest in Britain is planted up with money from these polluting companies, is this not progress?[6]

The answer is yes of course, if there is no other way of paying for the natural sequestration. But considering the polluter-pays and public goods principles above shows us that, for countries, there are other options, and the role of company sequestration needs to be considered in this wider context.

One further complication arises with trees. As they grow they sequestrate carbon and store it. But what happens when they reach maturity? Is the carbon stored in perpetuity or released? If they are cut down and the wood used for construction, the carbon store remains. But if the wood is burnt in a biomass plant, then at this point, having paid for the sequestration, the carbon tax should now apply. Drax, for example, should be taxed on its biomass-burning emissions, just as it should be taxed for burning coal. Even in the timber construction case, some wood will be wasted and again it matters whether it is burnt, stored or buried.

Offsetting for countries

A country is not like a company. It raises taxes, provides public goods and sets overall carbon targets and the associated policies. It considers the emissions in aggregate to hit its aggregate net zero target. The companies and individuals in the country will go on emitting even at net zero. What is it going to do to make sure that the economy as a whole offsets these so that there is net gain?

Governments have lots of options that companies and individuals do not. There are three possibilities: using the carbon taxes; creating new green infrastructures; and buying international credits. A carbon tax is the obvious way of imposing the polluter-pays principle, and hence improving the efficiency of the economy by internalising all the costs of economic activities. Polluters, faced with the consequences of their actions, will reduce their emissions. In the example above, the oil company might be paying for the natural sequestration through tree planting to offset the carbon tax it would have paid for the emissions had it not offset them.

If the company had instead paid the carbon tax, the government could use that revenue to plant the trees. The carbon tax could, for example, fund the Northern Forest. Natural sequestration would take place, just as it would have done had the company paid directly for these trees to be planted. In the economy as a whole there is an offset that helps to meet the overall aggregate net zero emissions target. It is just that it is the government and not the company that has the objective and the legal duty to meet it. Requiring each and every company to offset its emissions is neither necessary nor desirable if each and every company is already paying through the carbon tax for the pollution it is causing.

The government has a further advantage in investing in natural sequestration. A company will struggle to capture the benefits of all the *other* environmental gains from the tree planting. The polluter might have the choice between planting fast-growing spruce trees in dense plantations, or even eucalyptus trees, both of which are excellent at fast carbon capture. This would be a good option if all it wanted was to be net zero. These trees, however, have a considerable downside. The dense conifer forests tend to acidify the water, affect the soil pH, and do little to encourage biodiversity.

Suppose instead a mixed woodland was planted, with clearings. This would have much more biodiversity not only than the conifer forest but also (very likely) than the land on which it is planted. Planting in the headwaters of rivers prone to flooding would help in absorbing water and hence mitigating flooding, and might even directly improve the water quality too. A broadleaf forest with clearings might be accessible for recreation and thereby bring physical and mental health benefits. Different trees have different impacts on air quality.[7]

While the polluting company will not be able to monetise many or indeed any of these additional benefits, and will be disadvantaged by planting trees with lower sequestration rates and with higher costs, for a country these benefits can be captured in a host of ways. The flood defences will be cheaper; the costs of water treatment may fall; healthcare services might have lower costs from treating obesity and mental illness; and biodiversity is a clear public good. Government could have a comprehensive natural sequestration policy as part of its efforts to improve the multiple impacts on the environment. It could re-green cities, partly for other pollution and health reasons, create green walls, and plant trees along every street, all with the additional benefits of natural carbon sequestration.

In this calculation of the relative merits of companies versus the government doing the offsetting, and in the presence of an appropriate carbon tax, the government wins. The problem is that there is not at present an appropriate carbon tax, and governments are not internalising all these opportunities. The second-best world where governments choose seriously suboptimal policies leaves open a rather less attractive but nevertheless practical choice: between the companies doing the offsetting and the offsetting not happening at all, and with the companies over-producing carbon emissions because as polluters they are not paying the costs. It would be better to develop and credibly implement a wider environmental policy which captures these other gains. In England, the 25 Year Environment Plan aims to do exactly that, but whether it is fully and properly implemented, and over what timescale, remains to be seen.[8]

With the carbon tax revenues to hand, the second option for government is to use a portion of these to invest in green infrastructures, some of which would have the by-product of increasing natural sequestration. It could also invest directly in the CCS networks.

The main green infrastructure is the country's natural capital. If the government aims to maintain and enhance its natural capital, so as to leave the next generation with a stock of natural capital assets at least as good as it inherited, it is very likely that this will increase natural sequestration as a result, thereby offsetting some of the emissions.

With CCS, it has been argued that the government should directly subsidise at least one major demonstration project. But why not go further? Why not use the revenues from a carbon tax to build, own and operate CCS pipelines and a series of storage wells? A new company – let's call it British CCS – could receive monies from the carbon tax and invest in the sequestration. Instead of each company trying to offset its own emissions through CCS, the government could do it. The resulting activity would benefit from economies of scale and may be a natural monopoly. It could be a utility. The level of effective sequestration would be higher because the costs would be lower, including the cost of capital.

A country might try to offset the outstanding emissions over and above the net zero levels that remained by buying international credits. This is the third option. The UK government could add up UK emissions, see how far they exceed the carbon budget and buy emissions reductions overseas accordingly. This might be an efficient way forward: it could be cheaper to pay others to do the decarbonisation for it.

International carbon credits appeal to economists, who look for these least-cost opportunities. There is a global supply cost schedule of possible emissions reductions, and the optimal economic answer is to work down this curve for any given target. For example, it might be better to close a coal mine or a coal power station with a low thermal efficiency in China or India before closing Drax. It might be better to save the Amazon rainforest before planting trees at home.

This looks like a good way out of hard domestic choices, but a closer inspection of what is going on suggests why it may not be the best way forward.[9] Shouldn't that coal mine or coal power station close anyway? Shouldn't the Amazon be protected anyway? If the aim is to meet the 2°C target by, say, 2050, everyone is going to have to head for net zero. Either the unilateral net zero carbon target is genuinely unilateral, or it is global. If the former, then it has to be on consumption, not production, because otherwise UK net zero might actually increase the likelihood of the Chinese mine and coal plant staying open, and more of the Amazon being cut down, as we substitute home production for imports that may come from China or Brazil.

A unilateral target is aimed at reducing the UK's impact on climate change to zero. It is not designed to reduce some other country's emissions. If the aim is to address global warming in a multilateral way, the best place to start is with the low-hanging fruit, wherever it may be. But even in this case, it should be additional, over and above the reductions in domestic emissions, not instead of them. It is not strictly an offset.[10]

Net zero for individuals

Independently of the government or companies, any individual might want to offset emissions out of a moral duty to avoid being personally responsible for causing further climate change. How would you do this?

Recall your carbon diary. You will need to know what your emissions are before you can work out what you need to do to offset them. You need a baseline. You would then need to know which emissions you could stop causing on your own account, and those which are in practice unavoidable.

Let's start with mitigating your emissions. Analyse your carbon diary and see what the big-ticket items are first. There is the car and car travel. Do you need to emit so much carbon when you travel, and therefore offset? Or could you buy a smaller car, a hybrid or an electric car, or better still go by public transport, bike, or perhaps walk? The bit you might offset is the residual after you have made these adjustments to your travel arrangements. You may take one or more flights. Are these for holidays or for work? Could you go by train, holiday in the UK, or use conference and video calls instead? Some of this might save you money and make you healthier, although cheap flights are very different to high-cost rail journeys.

These are the easier bits to sort out and measure. The harder stuff relates to your consumption of food, drinks, clothes, entertainment services, water, heating, and so on. There is not much chance of you working this out precisely, so the net gain principle will apply and you will need to overcompensate.[11]

After the mitigations, there will be a lot of emissions you still cause. It is unlikely that you personally could do the offsetting, by planting trees and improving the soils, for example. If you have a garden, you could still make sure it does the best it can environmentally. You could make sure all the surfaces are green rather than concrete, cover the outside walls in plants, and so on. You might volunteer for the Woodland Trust. But lots of emissions will remain. Your airline may helpfully offer you the option of paying a bit more for an offset, and you could join the roughly 2 per cent of passengers who did this in 2019. But how can you know whether this really will be additional carbon offsetting, as opposed to corporate greenwash, and largely wasted? You should ask the airline what precisely your money is going on, how much of the money collected is actually spent on the sequestration, and how much on administration. Ryanair, for example, spent some of the revenue it generates from ticket sales on the protection of whales and dolphins in Irish waters, and as of 2019 it had one 7-hectare site for the planting of trees in Ireland.[12] You should ask this question of any charity you donate to, and this is no different. Suspicion may be justified.

The airline simplifies the process. It is relatively easy. You part with a small sum, and your conscience is clear, or at least you may think it is. It is more complicated for the other stuff you consume. You would need to make a rough calculation of how many tonnes of carbon you are responsible for emitting per annum, and then approach one of the many online platforms that offer to do this for you. Best to do a bit of research and find out what they actually do, but in principle you can do this.

Better still would be to have a proper carbon tax in place. Then you would be paying for the cost of the carbon you emit, and at net zero this will be equal to the sequestration necessary to offset your residual emissions. The Ryanair ticket in the example above would be much more expensive, and the government could then do some serious carbon sequestration. You could lobby your MP to get the government to impose a carbon tax to make you pay.

Offsetting markets

In the absence of a proper carbon tax, companies, countries and individuals can and should all offset the pollution they are not paying for. A central feature of these offsets is that the buyers of the offsets need to be matched to the sellers. Offset sellers need to be able to credibly convince the buyers that the proposed sequestration will actually happen. Market-making is a specialist economic function, and the credibility comes from the form and content of the offsetting contract.

Let's start with these market-makers and their incentives. They take money from the buyers of the offsets and then pay the sellers. The scope for corrupt practices is immediately apparent. Why not take the money and run? As with financial services, this obvious incentive leads to the need for regulation, and in particular the placing of these funds in safe forms of deposit, while they wait to be deployed. The funds should be regulated like any other financial service.

There are considerable portfolio and scale advantages. The size of the fund determines the size of the project. For a variety of reasons, large-scale projects may be better in terms of their overall economic effects than smaller ones, and the administration and other costs may be lower per unit of carbon saved. Large-scale projects often have higher proportionate associated benefits to biodiversity and some of the other co-benefits from natural capital. Bigger funds may, however, corner the market and charge excessive fees.

The incentive problems might be ameliorated by the institutional structure of the market-maker. For example, a charitable trust, without shareholders and dividends, might be more credible and less open to exploiting market power and corrupt practices than an ordinary company, which can go into liquidation, and has limited liability. The Woodland Trust is an example of a charity playing this role. Mutuals might, as in financial services, have advantages too.

A further reason for considering trusts is that the offset may take years to play out. As noted above, trees take time; they need capital maintenance and things may go wrong. It is important to know in advance what happens if things do indeed turn out differently and who would be responsible.

It is to be expected in the early days of voluntary offsetting that there are many players and lots of start-ups. As in any market, after an initial burst of activity, the survivors look for external validation of their performance. If you want to switch electricity supplier, you can look at the regulator's website and you know that it has vetted and licensed companies as a pre-condition for allowing them to operate. This sort of licensing and regulation has yet to emerge in the offsetting markets, but it should. This oversight is a public good for all who rely on the offsets.

A great opportunity

Sequestration is the poor relation of emissions reductions in climate change strategies. It shouldn't be. Done naturally, as part of a comprehensive environmental policy, it has the potential not only to make net zero an achievable goal, but to do this in a way that produces multiple other benefits. It can be a vital part of decarbonisation, and at the same time deliver a green and prosperous land. In doing this it helps to address not only the emissions head-on, but also ameliorates some of the damage climate change will do. New wetlands and coastal marshes help to absorb tidal shocks from higher sea levels and storm surges, and sequestrate and store a lot of carbon. Sequestration of carbon in the soil gives a big leg-up to biodiversity. Planting trees along every urban street sucks up pollution, improves our physical and mental well-being, and gives urban nature a helping hand. These sorts of intervention are close to no regrets.

The other great thing that sequestration brings is that it is something that individuals, as the ultimate polluters, can do. You and I can offset, and so can companies and governments. We can all just get on with it, especially in the absence of a proper carbon tax.

Adding natural sequestration and CCS to the suite of options gets us a long way down the road to addressing carbon consumption, and hence to net zero. Together with this green infrastructure added to the rest of the infrastructure needed for decarbonisation, and with a carbon price, the key tools are in place. They can now be applied to agriculture, transport and energy.

PART THREE

Agriculture, Transport and Electricity

8

AGRICULTURE: GREEN, PROSPEROUS AND LOW-CARBON

Most environmental problems come back to agriculture, and climate change is no exception. Climate change cannot be cracked without a fundamental change in agriculture. It is not just that agriculture is responsible for around a quarter of all global emissions, but that the way the land is used determines much of the other side of the carbon equation, natural sequestration. The atmosphere is a delicate balance of emissions and sequestration, and it therefore depends on what happens in the soils and vegetation, as well of course as in the oceans.

Modern agriculture (and aquaculture) does a lot of emitting. In the quest for ever more productivity from the land, the Haber-Bosch process for producing artificial fertilisers stands out as one of the greatest advances. By manufacturing fertiliser, the old dependency on animal manure and guano (essentially bird droppings) was broken, and with it the mixed-farming model. Add in pesticides and herbicides, and the conditions were set for intensive monocultures. The soils became blotting paper for the chemical mixes that the agrichemical industry kept inventing. Agriculture has always been a fight against nature, and chemistry upped the game dramatically.

The result is decimating to biodiversity. In Britain 97 per cent of wildflower meadows were lost between the 1930s and mid-1980s, and more than half the insects too.[1] This is all well known, and environmentally disastrous. What is less recognised is the stripping of the carbon from the soils, and the scale of agriculture's fossil fuel dependency, which in parallel has contributed to this collapse of biodiversity. Worse still, farmers are even being subsidised to encourage the direct use of fossil fuels: agricultural 'red' diesel in the UK is less than half the retail price of regular 'white' diesel, and almost everywhere globally governments subsidise farmers to do bad things to the environment in the pursuit of higher output and rural votes.

As this realisation has finally begun to dawn, the CCC has concluded that agricultural practices will have to change dramatically in the UK to meet the unilateral net zero target.[2] Its proposals are dramatic: 20 per cent of the land should be reforested, and meat production should fall and lower-meat diets encouraged. The NFU, attracted by the prospect of public goods and carbon subsidies, wants British agriculture to be net zero by 2040.[3]

While the CCC is belatedly right to turn the spotlight on agriculture, its initial proposals are simplistic. The future of a sustainable agriculture consistent with, and contributing to, the net zero target needs a lot more thought. The risk is that its two key proposals - more forestry and less meat - provide a narrow carbon focus without considering the wider ramifications, and could even become counterproductive.[4]

The baseline

In the transition to net zero, it is always a good idea to be crystal-clear about the baseline. It is to this that the changes have to be applied, not some hypothetical optimal model. In the case of British agriculture, the opportunities are great because the baseline is so bad: chronically inefficient; overwhelmingly dependent on subsidies; and with high levels of pollution for which it pays little or nothing.

The crazy economics of farming, described at length in my book *Green and Prosperous Land*, are fairly simple to state. British farmers annually produce around £9 billion in output (and not the £100 billion-plus number that the NFU keeps quoting for the food industry as a whole). This is around 0.6 per cent of GDP, and a falling percentage as GDP rises. The £9 billion is probably an overestimate because it includes lots of things that should not really be called agricultural outputs (like grouse and pheasant shoots, and maize and other crops for anaerobic digesters).

This very small industry is less important in economic terms than even these numbers suggest, since around £3 billion come from the EU's CAP subsidies and is guaranteed for several years to come post-Brexit. Farmers receive lots of other subsidies too, from the red diesel mentioned above, to special treatment under the inheritance tax system, exemption from business rates, easier planning regulations, R&D grants, and payments for losing livestock to diseases such as tuberculosis, foot-and-mouth and BSE (bovine spongiform encephalopathy).

To all these subsidies, worth perhaps half of the total output, should be added the costs of the pollution agriculture causes. British agriculture pollutes the waterways and much of the marine environment with fertiliser and pesticide runoff, and silt in the ditches and rivers from runoff due to exposed soils. It pollutes both the water and the land through the excrement of animals fed antibiotics and a host of pharmaceuticals to tackle worms and other internal parasites, diseases and foot rot. The natural environment now has to deal with a range of escaped aliens. The riverbanks and road verges now have lots of oilseed rape, for example, and the high levels of nitrogen lead to the destruction of much plant biodiversity.

Now add in the carbon emissions. Fertilisers and other agrichemicals comprise some of the most energy-intensive, and still carbon-intensive, products in the world. The diesel engine pervades farming, as tractors, combines and large vehicles have replaced horse and manpower. They are the power stations of the farmed landscapes. Dryers deal with the moisture in corn, and transport is a big-ticket item. Even seasonal labour flies back and forth from places like Romania and Bulgaria. The application of these chemicals and machinery has resulted in the large-scale loss of carbon from the soils, and the removal of miles of hedgerows as natural carbon sinks. Finally, there are the petrochemicals. Agriculture is now a plastic-dependent industry.

Add all this lot together and it is questionable whether British agriculture has a positive net economic output at all. This matters, not to belittle the efforts of farmers, but to recognise that the economic costs of changing agricultural practices in the pursuit of net zero are very low. It is one of the cheapest options available to us, compared with other big emitters like transport. It is therefore surprising that it has taken until 2019 and the CCC's net zero report for this to be recognised, and only because the scale of the decarbonisation means that we need to move on from just the electricity sector.[5]

The baseline is an agriculture which is inefficient in the most basic terms. Most of the measures to reduce emissions and increase natural carbon sequestration in agriculture improve economic efficiency, if only because it is pretty difficult to make it more inefficient than it already is.

It is not just inefficient. It is also largely uncompetitive. Imagine if there were no subsidies and no tariffs between the EU (under which the UK's agriculture has sheltered) and the rest of the world. Imagine a world of 'free' trade, where both subsidies and tariffs to protect British agriculture were abolished.[6] How much would survive? The answer is not much, and in particular virtually none of the cattle and sheep production grazed on grass pastures that make up most of the farmland. Britain is a small, crowded island with a jumble of rocks in one of the most heterogeneous geologies in the world. The Pennines, the Welsh mountains, the moors of the south-west, and much of the rest of the western side of Britain are not suitable for cereals, and the landscape with its small fields and hedgerows does not lend itself to the application of the massive machinery seen in the Ukraine and the American Midwest. Take a look at the pictures of lines of combines simultaneously harvesting wheat on vast farms overseas, and compare with the scale and size of British fields and farms. Even when it comes to livestock, compare pictures of Argentina's Pampas beef ranching, New Zealand's extensive sheep farms, or the corn-fed cattle lots of Texas.

It is hard to conclude that UK agriculture would survive a genuinely competitive global market, even if adjustments are made for animal welfare standards. Indeed, even with the massive subsidies, British agriculture can claim to meet only around 60 per cent of UK food demand (and this number is less than it seems). Go back to your carbon diary and look at what you eat. How much of it comes from overseas? Think of the avocados, the palm oil in virtually everything, the olive oil, and the fresh beans and strawberries all year round. The modern food supply chain is immensely complex, and few products that you buy are entirely local; even when they are local, they are not necessarily lower-carbon than some imports. 'Food miles' is not simply a matter of measuring the distance a product has travelled.

It would get quite a lot worse under a net zero unilateral carbon production target. Once a carbon price is applied to British agriculture, overseas producers will have another big advantage. This is equivalent to an import subsidy. Agriculture is one key area where carbon consumption really matters. In order to ensure that the impact on climate change is reduced to zero by 2050, imported food should be treated on the same carbon basis as domestic production. That means that 'cheap' food will not be as cheap as many assume. Like the political desire for 'cheap' energy which is also clean, 'cheap' food has little relation to decarbonised food. Net zero carbon consumption means higher, not lower, food prices – and in some cases much higher.

Polluter pays

It does not have to be like this. It is perfectly possible to have a sustainable agriculture which meets food production requirements, enhances biodiversity, and is consistent with net zero carbon consumption. It can be green, prosperous and low-carbon, even if it is not cheap.

To see how this could be achieved, think of agriculture like any other industry. To be efficient, which includes being sustainable, all pollution costs need to be internalised; all the relevant public goods should be provided; and there should be compensation for any damage to the stock of natural capital assets.

The important point about applying the polluter-pays principle is that agriculture has *multiple* pollutants. It is not just about carbon, and indeed if only the carbon pollution is priced there are going to be some unintended consequences, many of which could well be negative. The carbon price on emissions is necessary but not sufficient. It should be set at the same level as the rest of the economy, and applied to food imports too. This ensures that the cheapest emission reductions are done first, and these may well be in agriculture ahead of transport and even some electricity generation. The common price means that the same is the case for sequestration. Trees and soil carbon capture may be cheaper than CCS.

Put another way, if the carbon price is not applied to agriculture, is set at a different level from the rest of the economy, or does not apply to imported food, then emissions will be suboptimal. If there is no price, it is highly unlikely that the net zero target could be achieved. Although in theory it could all be done by direct subsidy, in practice, both the level of the subsidy and the way that lobbying and capture by vested interests work would make it all but impossible to gain the equivalent reductions that the common carbon price would achieve. Just think what a lobbying bonanza the NFU would be presented with if the government said it was going for net zero for the economy, more than net zero from agriculture (to compensate for the harder and more expensive decarbonisation elsewhere), and was going to pay for it all in direct subsidies to farmers. No wonder the NFU wants to be net zero by 2040 if it is going to be paid to do it, assuming that the polluted (the taxpayers), and not the polluters, pay.

The other pollutants need to be incorporated for agriculture to be sustainable and for the decarbonisation to respect the other environmental concerns. Fertilisers not only pollute the atmosphere, but the runoff also pollutes the rivers, and pesticides seep into groundwater. Agrichemicals harm wildlife and their application damages the biodiversity in the soils. Salmon farms encourage sea lice and pollute seabeds and freshwater lochs.[7] Unlike carbon emissions, these other pollutants tend to be diffuse and are location-dependent. A biodiversity premium can be added to the general prices of these inputs and, as suggested in *Green and Prosperous Land*, can be paid into a Nature Fund which would then ensure net gain. More on this later.

The most obvious area where the different pollutants interact is forestry. The CCC recommends that just under 20 per cent of farmland is reforested as a carbon sink by 2050. The obvious way to pay for this is via a carbon price. But if there is only a carbon price and no other intervention to deal with the biodiversity dimension, the incentive on landowners is to plant the trees that sequestrate carbon fastest. These are almost certainly non-native species with limited biodiversity gains. And they might be planted on peat too, doing further damage. If there is a price for carbon *and* a price for biodiversity gain, the incentive is to jointly maximise across both. Add in a price for the impacts on water flows, flooding and water quality. The optimal strategy is a *combination* of internalising all these externalities, not just carbon. A dense spruce forest might be best for the carbon, but terrible for all the other dimensions.

Agricultural public goods

With prices (taxes) on pollution, and prices (subsidies) on sequestration, agriculture will be a lot more sustainable. Applying the polluter-pays principle would revolutionise agriculture and the way the land is managed. But it still would not be sustainable. It also requires that the public goods are provided. Recall that these are goods and services that would not otherwise be provided because they are non-rival and non-excludable, and tend to have zero marginal costs.

The 25 Year Environment Plan published in 2018, and the subsequent Agriculture Act, put the principle of 'public money for public goods' at the heart of the subsidy regime. Lobbyists have responded by trying to prove that what they currently do is a public good. The classic move has again come from the master lobbyists, the NFU. Some of its leaders have argued that food itself is a public good, and furthermore that food security is too. Some NFU members in parliament and government, including ministers, a chair of the relevant select committee that oversees Defra (Department for Environment, Food and Rural Affairs), and numerous influential NFU members of the House of Lords, have been quick to exploit this, and thereby blur the distinction between the interests of farmers, public goods and the separate concept of public interest.

Food is not a public good: it is rival and excludable, and hence a quintessential private good. Food security is superficially a more plausible case, in that security itself is for all, like defence. The problem is in sorting out why the food bit needs security, any more than steel, cement or chocolate, and why this would not be provided, as it is every day in the supermarket, by private markets. The NFU frequently blurs the distinction between food and agriculture, whereas much of what constitutes the food we buy combines a host of economic activities, from processing and enhancements to transport and packaging.

It is true that in a war security is a public good for almost everything that is needed to survive and drive a war economy, and it is particularly true that food was a key element of national survival in World War II. But that was 70 years ago. In a modern war, with cyber attacks and chemical and nuclear weapons, there will be little chance of lasting long enough to be starved out under a long siege. The Covid-19 lockdowns revealed that it is the food supply chains, not the domestic food production, where security problems arise, and these in fact proved surprisingly resilient to such a difficult and unanticipated shock.

The public goods that have a bearing on carbon are the more fundamental ecological systems on which agriculture depends, and the natural sequestration of carbon. These systems are made up of natural capital assets, and for agriculture they include the water, air and soils within which biodiversity is embedded. Since soil carbon is closely correlated with biodiversity, and since most biodiversity is beneath our feet, it can be argued that soil has public good aspects.

Sustainable agriculture needs to respect the intergenerational dimensions, and in particular the stewardship duty to pass on to the next generation a set of natural assets that are at least as good as the ones the current generation inherited.[8] This natural capital is the infrastructure upon which the future is built, and the carbon within the natural capital and the ability of these assets to sequestrate are the natural capital systems that sustain life more generally. The damage to soils will be addressed in part through the carbon price, but the additional biodiversity interests will need further support, and it is here that subsidy should be concentrated.

As with all the other sectors in the economy, the ability to contribute to net zero depends on technology and, in particular, the development of new technologies. As noted, R&D is as close to a pure public good as it gets. R&D in genetics, gene editing, genetic engineering, the carbon sequestration potential of different trees and plants, soil science and knowledge about the complexities of biodiversity are all public goods, as is the climate modelling that the IPCC has helped to drive forward. Scientists in publicly funded laboratories contribute to this public good. A sustainable agriculture has a part in science funding generally.

Dealing with the damage: net gain

The final hurdle that a sustainable agriculture has to clear is where it continues to do damage despite the provision of public goods and prices on its multiple pollutants. Agriculture will always modify the land on which it depends, and there will always be a trade-off between maximising crop yields and maximising biodiversity, water quality, and so on. In agriculture's long fight against nature, this tension will not be entirely abolished by net zero policies. It is the way the balance is struck that should and will change, including for net zero carbon reasons.

The aim of the polluter-pays principle is to internalise the costs that agriculture imposes. The damage will have to be paid for, and applying the principle will reduce the damage. It will not, however, eliminate it. Sustainable agriculture is not the same as maximising nature. Farming in harmony with perfect nature is a demand too far. If it is nature that is the sole concern, a great deal of agriculture would simply cease, and alternative land management approaches would be taken. This is indeed what many 'rewilders' aspire to.

Globally, land drainage, land clearance, hedgerow removals and a host of measures which increase yields will continue to take place. Most of these steps have been taken in Britain and Europe, but it is a different story in developing countries. Think of the rainforest clearances for palm oil and cattle ranching. Even with the application of the polluter-pays principle, and the provisions of public goods, incentives will remain in numerous cases to continue to do damage to natural capital. It is for this practical reason that the polluter-pays principle needs to be buttressed in agriculture by net gain. Where damage is done, it should be compensated for, over and above the carbon and other prices. This applies to imports as well as UK production.

This not only ensures that, for example, palm oil producers have to take seriously the permanent loss to the ecosystems and their natural capital assets that they cause, but it also helps to check the incorporating of agricultural land into developments and urbanisations. When greenbelt land is concreted over, it will probably never return to its previous uses. While it is often the case that what is called greenbelt is not actually green at all, it could be, providing the multiple natural capital services close to people, including local food. It could do a lot more natural carbon sequestration.

The net gain principle has this further twist. It is not just the cost–benefit comparison of what the land is used for now with the alternatives, but it is also the benefits that natural capital yields *for ever*, provided it is not driven to such a low level that it cannot replicate itself going forward. This is why the net gain principle is a critical part of the broader demand to be precautionary. Eliminating a species is *for ever*, and we do not know all its future potential benefits. The permanent loss of a natural asset is a much greater risk to take than merely the loss today.

Agriculture with these three principles applied would be very different from that which exists today. It would be one with far fewer carbon emissions; the land would be an integrated part of the carbon cycle; and natural sequestration would be an intimate part of the way farmers manage their land. The public goods would be provided, and natural capital as a whole would be protected and enhanced, in perpetuity, fulfilling intergenerational stewardship duties. It would be farming for carbon, for biodiversity and for food. It would also be much more economically efficient.

The new technologies, digitalisation, genetics and indoor farming

In doing all this, it turns out that there is another big benefit, which comes from encouraging further the technical change that is already well under way. Technological change provides a great opportunity to decarbonise agriculture *and* to free up the land to a much more carbon- and nature-friendly future. It is every bit as important and exciting as technical change in transport and electricity.

Farming has been subject to one technological revolution after another. The landscapes today are the product of artificial fertilisers, tractors, antibiotics, pesticides and genetic selection. Of these, it has been fossil fuels that have made the biggest difference, including enabling the large-scale production of fertilisers, and ushering in the tractor and other heavy machinery.

This last big fossil fuel technological revolution at the end of the nineteenth century turned farming from a labour- to a capital-intensive business. Human labour and horsepower gave way to the diesel engine. Unsurprisingly, the result has been massive emissions of carbon.

In the twenty-first century, another technological revolution has been building, this time based on two general-purpose breakthroughs. The first is digitalisation, and the second is in the life sciences and genetics. The two are of course related: without the new data technologies, genetics would be tough, relying on traditional techniques of animal and plant breeding, and the science of Mendel's peas.[9] The genetics in turn tells us how the data should be addressed.

Having become an energy-intensive industry with the coming of diesel engines and artificial fertilisers, the arrival of the internet, and the possibility of digitalising almost anything, agriculture has been opened up to becoming a data business too. The new information baselines for every square metre of land, the underlying soil characteristics, and weather and climate models enable farming to become a precision activity, relying on computers rather than the experience of farmers built over generations. Agriculture is energy plus data, to which water and soil are added.

Driving the farming activities, based on the data analysis, are increasingly autonomous vehicles. Robots are gradually taking over most of the rest of the manual farming activities, from milking cows to picking fruit. Lots of farming will soon be done remotely from computer screens relying on multiple sensors. There will be 'hands-free' fields.[10] The world within which the current farmers honed their expertise will be largely replaced by a new digital one. With the average age of farmers in most developed countries (including the US) at around 60, it will be a generational change too, with maths graduates increasingly taking over from students at agricultural training colleges.

Digitalisation is not confined to robots, data and AI, but also includes the things that these will enable. 3D printing ushers in a much greater degree of customisation, as products are printed from digital images and instructions. The food chain itself has already been digitalised, but there is more to come here too, with blockchain and the internet-of-things allowing precision identification and tracking of each item and payment. Agricultural activities are increasingly being integrated with food processing and logistics supply chains.

Food itself is being broken down by genetics. Where once there were cows and sheep, and then specific sorts of cows and sheep, it is increasingly possible to regard each as bunches of genes. Flowing on from the mapping-out of the human genome, this technology now expands into the possible genome of every living thing. It is an obvious next step to try to move on from old processes of selection by breeders, which have given us most of our agricultural species, to modifying them.

Two types of genetic management have emerged: gene editing and genetic modification. One version of the first is CRISPR (clustered regularly interspaced short palindromic repeats), and the second goes under the general heading of GMOs (genetically modified organisms).[11] The former is already part of medicine and plant and animal science. If a faulty gene is discovered, the 'editor' switches it off or cuts it out. It takes what is already in place and modifies it. There is, however, a world of difference between editing the genes of *one* individual, and editing the genes that are passed on to subsequent generations.

Gene editing differs from gene modification in that the latter imports genes from other plants and animals. It is altogether more controversial since some of the resulting plants and animals are different from the existing ones. It modifies species and creates something new. Gene editing takes what we have got and tries to make it work better. Gene modification creates new genetic sequences.

Gene modification is often challenged in the dramatic terms of 'playing God with nature'. It gives rise to fears of Frankenstein: that, once created, new modified organisms will escape into the wider environment. They might reproduce with unintended consequences. Scientists working in these areas are quick to try to reassure us. They will not be able to breed and will be strictly controlled. Despite this guarantee, GMOs have been banned in Europe, but not in the US.

From the carbon perspective there is a great opportunity here. The polluting aspects of farmed animals might be modified, and in particular their methane-producing characteristics may be altered. Plants could be modified to be better at sequestrating carbon as well as being used to produce new foods that replicate the proteins and other benefits of meat.[12] All this, and higher crop yields too. This may be much more successful and on a larger scale than some of the geoengineering proposals being developed to try to head off major climate change, should it materialise. The science cannot be put back in the box, and existing technologies cannot produce enough food for the bigger global populations. There will be 10 billion people to feed while also reaching net zero. It is not, and cannot be, one or the other.

Indoor farming offers up complementary possibilities, and quite a lot of food production takes place indoors already. Indoor animal rearing gives lots of opportunities for food and disease management. But it also risks harm to animal welfare. Sentient creatures, deprived of their natural habitats and the ability to benefit from the outside air, may suffer. The examples of battery hens packed into sheds, and pigs kept in pens on slats, have become unacceptable.

Sentient animals raise these issues with a sharp and potentially painful focus. It is less clear with fish and insects. GMO salmon can produce a lot of protein in concentrated tanks. They do not have to have anything to do with the sea or freshwater lochs. Indeed, it might be much better if they did not. Should we worry that they are modified? Might they escape and breed with wild fish? Not if completely separate in fish factories.

The potential of insect protein is one of the great as yet untapped ways of producing a lot more in factories. Insect protein is already fed to fish and added to animal feed. It could be put into human foods too, directly or as an additive. To give some idea of the scope and scale, a small insect factory could produce more protein than all the cattle and sheep on some of the largest farming estates.[13]

Indoor farming offers other benefits. Its emissions can be very strictly controlled. In theory it can be a closed system, with air and water pollution sealed within the unit. The management of animal slurry is one example, but there are opportunities of methane and carbon emissions abatement. For plants, because nutrients are very targeted, the use of fertilisers can be reduced, and pesticides are not needed at all in completely sealed units. In being pesticide-free, indoor farming could be regarded as organic.[14]

This leads to the next potential carbon saving, by freeing up some of the land from food production. On an indoor farm, it does not matter where it is produced. The required levels of light, water and nutrients can be provided irrespective of location. That means we can contemplate big carbon reductions from minimising the amount of transport in the food supply chains. Food production can be close to food consumption. This in turn may also help to control food waste, and hence require less production for any given level of consumption. This is the era of urban farms, farms in the desert, and farms in developing countries closer to their growing populations.

All of this is helpful to decarbonisation only if it itself is a low-carbon process. Indoor crop growing requires light, water and nutrients. The light comes from LEDs. LEDs require electricity, and to help decarbonise this needs to come from low-carbon sources. The LED lights and the indoor production processes need lots of renewables to generate the electricity and power the motors in the factories without diesel tractors. Indoor LED-based farming makes agriculture overwhelmingly an energy business, plus data. In Iceland, there is abundant geothermal energy, already allowing the cultivation of lots of salads, vegetables and tomatoes in greenhouses, lighting up the Arctic landscape. Elsewhere it has to be down to solar, wind and nuclear. Solar can do some of this, and plants do not need LED lighting 24 hours per day. In the Middle East the challenge is cooling, and all these new techniques will benefit if there is some electricity storage.

For many countries the problem is water, especially for livestock. Plants can grow in recycled systems, but animals need a constant supply of drinking water. This has historically been a serious issue in many countries. It is likely to get worse, as irrigation is often a temporary process, resulting in higher levels of salt and eventually abandonment of land. Climate change will alter the distribution of rainfall. Many countries are damming rivers, causing localised problems. The Tigris, the Euphrates and especially the Mekong are all examples of where this can go, and has gone, horribly wrong.

With current agricultural methods, it is not hard to paint a picture of running out of water to grow enough food to feed expanding populations. It is already a significant constraint. Yet the world is not short of water. Most of the surface of the planet is water. It is just that it is salt water. This raises the prospect – indeed, probably the necessity – of major desalination. As explained in chapter 6, the problems with desalination are that it uses a lot of energy, and the membranes between salt and fresh water are not that efficient. But what if solar technologies advance? What if more of the light spectrum is opened up and the development of solar films makes this energy much easier to extract? What if new materials like graphene radically improve osmosis? With effectively infinite solar energy supplies, desalination and LED lighting could become the core building blocks of future agriculture. Land could be freed up for other things, including natural sequestration.

A unified framework: a 25 Year Plan

Agriculture has to decarbonise to meet net zero. The diesel engine, the carbon-intensive artificial fertilisers, and the high-methane animal production all have to be dealt with. More of the land has to be devoted to natural sequestration, with many more trees and a return to high carbon concentrations in the soils. The peat bogs have to be protected, and agriculture cannot keep cutting down the great rainforests. Consumers have to pay for their carbon consumption whether goods are produced at home or abroad.

All of this spells massive change for agriculture. Changes in consumption and new technologies are a necessity, not an option. Freeing up the land for natural sequestration is possible only if more intensive indoor agriculture is facilitated. Indoor agriculture leaves the great outdoors to focus more on the public goods, with carbon sequestration and promoting biodiversity among the good things to do with the land.

This needs a framework which integrates the various externalities and public goods. A 25-year environment plan does this, and spans most of the period between now and the 2050 target. Every country needs such a plan. The version in England sets out 10 goals, from clean air, to thriving wildlife, good water quality and enhancing biosecurity. It is far from perfect, but it is at least roughly right.[15]

Making this framework deliver decarbonisation, while respecting other public goods, requires a common approach to each component. Carbon needs a common price across all the agricultural domains. The public good of biodiversity needs to be subsidised. Clean air cannot be simply a matter of displacing power stations away from cities (as in China). It must be accompanied by decarbonisation at the same time. All of this needs to accommodate a considerable increase in the demand for electricity, for a digitalised, desalinated and LED world.

None of this will be achieved by treating carbon as a problem separate from food production. To repeat, decarbonisation has to take place in the context of possibly 10 billion or more people, all of whom need food and energy. Demand for both food and electricity is going to increase a lot, creating a moving target which we are in danger of losing sight of. Simply decarbonising the existing economy will not meet the needs of all these extra people. Your carbon diary will have to change too: less meat, fewer imports, less food waste.

9

REINVENTING TRANSPORT

Like agriculture, transport is responsible for a big chunk of global carbon emissions and, like agriculture, tackling this can't be done without new technologies and changes in the way we travel.

Technical revolutions have happened several times in the past. We got from walking to wheels, and from coaches and horses to trains, cars and lorries, and from sail to modern diesel-powered shipping, and we worked out how to fly. Rockets can even transport us to the moon and beyond. Now we will have not only to find new, low-carbon ways of getting ourselves and goods around, but we will personally have to think hard about our own transport carbon footprint. Again, as with agriculture, net zero requires both new technologies and reductions in personal carbon consumption. Take a look at your carbon diary: do you really need to travel so much?

What the internal combustion engine did for us

Alongside the steam engine, and steam power stations, the invention that transformed our economic prospects, made the stupendous economic growth of the twentieth century possible, and caused many of our pollution problems today is the internal combustion engine and, above all, the diesel version.

This engine allowed the high-density energy potential of oil to create a transport network at both the very local and the global level. Getting anywhere before the internal combustion engine was a major undertaking. On land, it meant by foot, by horse and, from the mid-nineteenth century, by steam train. Before the railways, getting from London to Edinburgh took days, and getting from London to tour the sites of Europe, and especially Greece and Italy, took weeks. Whether it be Samuel Johnson's tour of Scotland or the 'grand tours' of Europe for the elites, diaries and journals tell these stories.[1] Travel was as much an achievement in itself as a means to an end.

Diesel replaced steam engines for shipping and eventually for trains. The current global network of trade would be very different if it still relied on sail and coal-fuelled steam engines. Cars and lorries opened up trading networks which had previously been restricted to the more limited rail and canal networks. Over the course of one century, the world went from cars as a luxury to having more than a billion of them on the roads. The Model T Ford opened up mass production, and that mass production needed lorries to make it work.

The only serious further technological development in the twentieth century to add to these nineteenth-century advances was the jet engine, and this also required lots of fossil fuels.

It is all about oil (and a bit of gas)

The oil industry exists to do two things: to facilitate transport and make petrochemicals. The former still dominates the latter. The early oil industries in Baku in what is now Azerbaijan, and Pennsylvania in the US, produced limited amounts of kerosene mainly for street lighting. (The biggest market was London.) But with the internal combustion engine beating the electric alternative in the late nineteenth century, the oil industry had an almost infinite market in front of it to supplement its supplies to ships, and especially the British Navy as it was converted to oil by Winston Churchill just before World War I.[2]

The march of the oil industry is also the march of carbon emissions. Early production primarily served local markets, but as oil transport by railroad picked up and then pipelines were developed, it could reach the parts other bulky fuels like coal could not, or at least not at competitive prices. Its dense concentrated energy potential made it a clear winner.

The staggering development of the oil industry is effectively the story of the great industrial expansion of the twentieth century. The twenty-first century has to bring this oil industry to a rapid end, and replace it with electricity and perhaps hydrogen, if there is to be any chance of cracking global warming. Almost everything that now relies on the internal combustion engine and the jet engine will have to be switched to something else (or at least to biofuels in the meantime). That, in turn, means a radical transformation of the car and vehicle industries, of the filling stations, of the electricity networks to support charging, and of batteries and storage. Decarbonisation needs a new transport industry.

The end of the internal combustion engine does not, however, automatically translate into an end for fossil fuels. Gas and coal still generate a lot of electricity. This is why the indirect demand for gas for electricity to power electric vehicles may hold up for decades to come. In an electric vehicle world, the big loser is oil. Gas has a possible role as a transition fuel; oil does not. Gas can also power LNG vehicles, again as part of a transition.

The electric alternative

Mitigating climate change in transport, as in agriculture, keeps coming back to electricity. In a net zero world there will be few vehicles powered by fossil fuels. Cars, lorries, ships and planes will all need to switch. And they will need to do all this before 2050 – in other words, in just one generation. It has to be much faster than the switch in transport to fossil fuels in the twentieth century. The only alternatives are biofuels (prolonging the life of the internal combustion engine) or hydrogen. The former has all sorts of problems as we shall see, while the latter will have to be produced by electricity in a net zero world.

Electric transport is, like its oil predecessor, about a whole supply chain and its supporting infrastructure. An electric car requires a battery that is light enough to make the vehicle driveable, and a charging network. Just as conventional cars need to be able to fill up their tanks, electric vehicles need to be able to plug into the electricity networks. This, in turn, means that there needs to be the supporting low-carbon electricity generation, the networks to carry it, and readily accessible charging points.

An electric car's battery is currently anything but climate-friendly.[3] Batteries are in effect chemistry sets. They require a number of core raw materials, and these have to be mined and are then subject to industrial processing so they can be added to the car battery. Mining elements such as cobalt is typically an environmental nightmare.

To these raw material challenges there are added all the other bits on the car. Where did the materials for the bodywork come from? Do they need steel, or plastics? How much energy goes into these production processes? What about the tyres and the brakes? These wear down and produce their own emissions.

Thinking about the full climate consequences of producing and driving electric cars quickly punctures the adverts which paint green driving scenarios in uncongested open countryside. There is nothing about wildflower meadows in the current electric cars. For this reason, some argue that the problem is not so much switching from fossil-fuelled cars to electric ones, but getting out of cars altogether.[4] [5] There is no such thing as a carbon-free car, and this is where your travel-related carbon consumption in your carbon diary comes in.

Batteries and charging networks

Put these other carbon pollution issues aside for a moment and concentrate on the batteries. There are two choices here. One is all about the battery technology, and specifically the lithium-ion battery versus other options. The second is about the standardisation versus heterogeneity of the chosen battery technology.

Lithium-ion is the battery of choice because it has had the most development, given its use in modern communications equipment.[6] There would be no laptops or mobile phones but for the miniaturisation of lithium-ion batteries. They were invented in the 1970s and patented by Sony in Japan in the 1980s, with more than 30 years of subsequent refinement. In the rush to win the manufacturing battle for electric vehicles, the appeal of a well-known and widely used battery technology over other options is obvious.

The good news is that lithium-ion does work, and at a battery size and weight that makes electric cars function. The better news is that there are lots of alternatives coming down the track that may be even better. If the future of electric cars depends on batteries, this is a problem that can be cracked, even if they will never be zero carbon, and hence sequestration will always be needed.

Batteries are about the temporary storage of electricity, and about replacing the temporary store of petrol or diesel in conventional fuel tanks. But whereas oil can be easily stored in tanks, and transported by pipeline and tanker vehicles, electricity needs wires and a supporting electricity transmission and distribution network.

Comprehensive networks for car battery charging do not exist. They will have to be created. Filling stations are not embedded in electricity networks for the offtake of significant power, and they are not designed for fast charging, with the consequent strong surges in demand. The national electricity transmission system and the local distribution networks were not built with filling stations in mind.

Electric vehicles thus become another infrastructure problem: enabling the charging of batteries. It is here that some really exciting opportunities emerge. Why take the electricity to the car rather than taking the battery to the electricity network supplies? Why charge batteries instantly, when the batteries can be swapped, and then taken away for recharging at a time and place which best utilises what we already have?

This presents two possibilities. The first is the one we are heading for. You drive onto a motorway services forecourt and plug in. You wait while the car charges. You want this to happen quickly and you may be willing to pay a premium for speed. It will take a bit longer than putting in the petrol or diesel, and you don't want to spend hours at the service station, so the quicker the charging the better. Then off you go again. It is as exact a replication of the current system, with some of its speed and convenience, as you will realistically get.

The second option is even quicker. It is like a Formula One motor racing pit stop for tyres. Racing driver Lewis Hamilton drives into the pits, off come the tyres, on go a new set, and in less than 10 seconds he is on his way again. Imagine if, instead of the tyres, the depleted battery is taken out and replaced with a charged one.[7] It could take a matter of seconds, and off you go out of the motorway service station. It would be bad news for the service station shops, just as it would be for airport shops if you could go straight through an airport and onto your plane, as there would be less time to sell you food, clothing and other stuff. But it would be really time-efficient and hassle-free, although the total stock of batteries would need to be larger. It could be done for trains too. Instead of all those ugly overhead electric cables, and the vulnerability of the railway networks to power cuts and overhead line damage, the train could have a 'pit stop' at key station intervals and batteries could be swapped at speed. It might well turn out to be much cheaper too.

Why doesn't this happen? Enter the car companies, competing for market share. What do they offer customers in an electric car market? Answer: they will claim that their battery is better, and better designed to fit in their vehicle and with a better weight dynamic. The last thing a competitive car manufacturer wants is for the customer to see its vehicle as the same as all the rest.

The car industry lobbyists will point to the problems of agreeing which battery design should be the one that is used. How big should it be, since the size would need to be uniform? Which technology? It is all very familiar. Oil companies used to claim that their petrol was superior to that of its rivals. 'Put a tiger in your tank' was a classic ESSO advertising line. But when uniformity was imposed, with a rating system up to four-star, nobody noticed any difference, because there wasn't much. Part of this was because some refineries supplied fuel to different brands. Grangemouth in Scotland under BP, for example, once supplied all the petrol in Scotland, regardless of the brand on the filling station forecourt.

It is easy to overstate how much uniformity is needed for a battery-swapping infrastructure. There could be a range of sizes, with the charging station stocking each: large, medium and small car versions, for example. The internals could vary a bit between the battery manufacturers. You could swap for a 'basic' or a 'special', with different battery lives before recharging, with discounts and all the various marketing gimmicks that companies love to promote.

The opportunity for battery swapping in commercial vehicles may be greater. Here, the marketing features for the manufacturers are more about economy and efficiency. There are some large van operators, including the main utilities and delivery services. Even if the market as a whole does not promote battery swapping for all vehicles, each large fleet operator could do it for themselves. Fast forward to car rental and urban car clubs and again there are fleet options.

All this matters a lot. In the battery-swapping case, the time of the day or night when the batteries are charged makes no difference to the driver. The batteries can be recharged whenever there is a system surplus, and in particular at night. The charging process exploits the variance in the other electricity system demands. In theory, no new power stations might be needed for the system.[8] It would just use the existing power stations more of the time.

It gets even better. In theory, the charging could be 'off grid' altogether. Since it does not matter when the batteries are charged, the intermittency of renewables would not matter. The batteries could be charged when the wind blows and during the day for solar panels.

The point is that the overall system costs might be remarkably low compared with the instant fast charging. The key obstacle is the classic one of getting agreement to pursue the public system good, rather than the narrow interests of each car company considered in isolation. This takes bold government action and planning, not a *laissez-faire* approach.

These two cases are deliberately stylised to throw up their very different costs. Is there any halfway house, leaving some choice but still keeping the system costs down? The obvious one is for each driver to behave as if the battery-swapping exercise takes place. The way to do this is for each to charge the car when the system is in surplus, and that means overnight. If electricity prices are very low at night, and the costs of fast charging during the day are very high, then there is a powerful incentive to charge at night.

This works provided that the mileage travelled during the day is not greater than the battery charge life. You go home in the evening, plug in your car, and expect it to be fully charged when you leave for work in the morning. Again, the downtime on the system is exploited, and the charging can be varied over the period from, say, 9pm to 7am by smart technology which optimises your charging against the system costs.

Note that it is still more expensive and you don't get an instantly charged battery. You have to wait, and this time the system needs to make sure that the charging points at your home (and those in the street) are able to cope and charge. It requires a major overhaul of the electricity distribution system. In the battery-swapping case, the batteries can be taken to a central depot and plugged into the National Grid to optimise the system further. To see the opportunity here, imagine if the National Grid owned the recharging depots, just as the oil companies own many of the filling stations. It would plan and enhance the grid to match the distribution of road use and ease the resupply chain for the batteries in line with its existing transmission lines.

It looks like we are going down the route of a mix of home charging and rapid charging, and the result will be a much greater cost to decarbonisation. Choice comes at a price. It might improve as autonomous vehicles come onto the road systems. If the car is autonomous and guided by smart systems, it may be that there is a shift in our relationship with the car, from the huge variety of styles, engines, colours, designs and interiors, towards seeing it as merely a way of getting from A to B. People rarely worry what type of car a taxi driver uses when summoning one, and when calling up an autonomous vehicle on an app, programming it to take you from A to B, you may not really care what sort of battery it has got. The market may respond to this. Suppose what matters to you for this A to B trip is the price. An autonomous vehicle provider (and not you) will own the car and will have powerful incentives to standardise the technologies of its fleet. It may see an advantage in battery swapping back at the autonomous car depot, or perhaps as part of the optimised journey timing and use.

This is what the car companies fear. They may not be able to extract the economic rents that go with advertising and branding. They may simply sell thousands of identical cars to a car pool company, with the capital provided by an infrastructure fund, and then the economic value comes in the convenience and frequency of the autonomous vehicle's availability, and not from packing it with all sorts of extras that the customer might not need. Why, for example, does the car need a sophisticated sound system when you can get it all directly via an app on your phone? Travellers do all this for themselves on the bus and the train.

The electricity network issues are addressed further in chapter 10.

Hydrogen and fuel cells

All the above assumes that the future is electric, and it almost certainly will be. Assuming this outcome makes it path-dependent and therefore inevitable. Making the investment now in batteries and electric vehicles will mean that this is the energy vector we end up using.

But this does not mean that each and every vehicle will be powered directly by electric batteries. There are other ways of powering vehicles from electricity. The prime one is hydrogen. Hydrogen can generate power without batteries, and produces only water in emissions. Cars could have fuel cells that turn the hydrogen into power for the vehicle. The key bit with hydrogen is not so much whether it can power vehicles (it can) or indeed whether it is practical (it can be), but how the hydrogen is made. There are two main ways: by reformulating natural gas, or by electrolysis and hence by electricity.

Oil and gas companies like the former since it gives them a new market and keeps them in the transport fuelling business. But it is not zero carbon. Gas is better than oil, but net zero and gas do not mix except at the margin and only with large-scale sequestration.

The hydrogen-from-electricity route may be expensive relative to using electricity directly in batteries, but there are some intriguing and special circumstances where the costs are very different. Imagine a remote location off the grid – say northern Norway which has a lot of wind potential, or the remote Canadian mountain areas with lots of hydro potential but far fewer markets in close proximity. Rather than building transmission networks for the electricity that could be produced in these remote locations, and bearing in mind the losses from long-distance transmission, might it not be better to turn that renewable energy into hydrogen, possibly via ammonia (and separately, to use this process to create low-carbon fertilisers too)?[9]

This matters because not only are there some sources of renewables remote from electricity grids, but there are also some forms of transport for which batteries and charging might not be feasible (in part because of recharging). The big one here is shipping, which currently is not only heavily carbon-polluting, but also uses some of the crudest forms of crude fuels, creating lots of sulphur emissions too. It also applies to long-distance, very large-scale vehicles.

The hydrogen problem, like that of electricity, is a storage one too. How can it be safely stored and then delivered to ships and large vehicles? This remains to be cracked, and whether and how it is will determine the batteries versus hydrogen competition. It is too early to arrive at any definitive solution.

The electrification of transport is a huge challenge, and it will not happen by accident. Nor will moral pressure do the job. While the educated and well-off might be willing to make the switch to electric vehicles, and a Tesla or electric BMW provide a status symbol for the better-off to display their carbon-savvy lifestyles, for the bulk of the population it is about costs and affordability. As long as petrol and diesel cars are cheaper to make, cheaper to buy and cheaper to run, and as long as they can be easily refuelled, then their hold on the market will continue, and for much longer that the net zero target demands. That is why SUV sales are booming.

What then to do? The answers for transport are the same as for agriculture. The three core principles need to be applied: the polluter should pay; the public goods need to be provided; and where pollution continues there needs to be net gain and hence offsetting. Together these incentivise innovation, R&D and a carbon consumption response.

Making the changes: start with the polluter pays and the carbon tax

The polluter-pays principle dictates a carbon price, applied both domestically and at the border. Domestically, petrol and diesel are already taxed very heavily. Indeed, once the full supply chain is considered, the energy tax already approximates what a carbon tax would look like in terms of levels.

This is how it works. The price of crude oil on the international market is much higher than the cost of the marginal barrel of oil produced from the cheapest sources. The cheapest is in Saudi Arabia and Iraq, where it can cost as little as US$5 a barrel (or less) to produce. Even in Russia the marginal cost is probably around $20. The market price is much higher and has, since the first oil price shock in the early 1970s, remained way above this marginal cost. At times, market prices have been over $100 a barrel, falling back in the late 2010s to around $60. While there are very good reasons to think this price will gradually fall if and when the world decarbonises (and the demand for oil starts to fall), it is going to be well above $5 for a very long time to come.[10]

Why is the price not equal to the marginal cost, as in other markets? Why doesn't normal economics work? The answer is that oil production is highly concentrated in three countries: Saudi Arabia, Russia and the US. Each produces around 10 mbd, out of a total market of around 100 mbd. Saudi Arabia and its OPEC allies can collude and thereby temporarily force the price up by cutting production, and this is in the interests of all the producers (including Russia and the US) – and perhaps even more effectively Saudi Arabia and Russia can rig the market as they did when the Covid-19 pandemic struck. In a normal competitive market, the cheapest reserves would be produced first. In this market, the Saudis in particular choose not to do this to their full potential capacity.

As the price is credibly kept well above $5 a barrel, other supplies come into the market which would not otherwise be produced. The North Sea oil and gas industry developed because of the oil shocks in the 1970s, with much higher costs than in the Middle East. So too did Alaska, the offshore Gulf of Mexico, and eventually even the tar sands of Alberta in Canada could turn a profit at the higher prices. These high-cost producers need the Saudis and others to manipulate the market to keep the price up.

The relevance of all this for carbon and climate change is that this premium on the costs of production has an effect similar to a carbon price. What might be called the OPEC Tax is the dominant carbon price in the world, and it applies universally across all countries, although some subsidise their consumers. A carbon tax would be *on top of* the world oil price. It is the difference between the marginal costs and the world oil price, plus the carbon price.

Looked at this way, it tells us that people are willing to carry on buying oil, even at the high implied tax rate that already applies. And it tells us that if the oil market was a normal competitive one, the price of oil would be much lower and hence demand would be correspondingly much higher – well over 100 mbd. Carbon pollution would be correspondingly much higher as a result. Raising the carbon price high enough to engineer a shift from fossil fuel cars to electric ones would need a further significant increase. Another way of putting this is that the overall competitive advantage of conventional cars is very significant.

The main initial impact of the carbon price for oil will be to encourage a switch *between* the fossil fuels, rather than from fossil fuel transport directly to electric. Of this shift, the biggest part is from oil to gas. Gas can fuel vehicles directly but, as noted, it can also generate the electricity for the electric cars and its costs remain for the time being well below those of many renewables and of nuclear. Gas has half the emissions of oil, so as a transitional step to decarbonisation it has some attractions. (As an aside, coal typically has no carbon price, and hence the substitution from coal- to gas-fired power stations is inefficiently held up in favour of coal.)

Carbon emissions are not the only source of transport pollution associated with fossil fuels. Burning oil in transport produces a cocktail of nasty stuff, including sulphur and acid rain, nitrogen oxides and particulates. All have consequences, and the polluter-pays principle dictates that all should be taxed. Most are not, but rather (typically inadequately) dealt with through regulation, notably car emission standards prevalent in the EU and the US, and limited access to city centres.

The problem with these other associated pollutants is that whereas location does not matter for carbon, it usually does for these. Location-specific taxes are notoriously difficult to set and adjust, even in urban areas, and so there is a good case for taking a regulatory route for many of these, plus local congestion taxes.

Transport is not just national, and it raises carbon consumption and hence carbon border adjustment issues. Think of global trade and hence shipping in energy-intensive products, like steel and fertilisers. These are transported by sea, before switching to the roads. If steel is produced in China rather than the UK, it not only has a lot of coal embedded in its production, but it also has all those shipping pollutants too. British Steel, supplying British markets, has less coal intensity in its production, and none of the shipping costs.[11] A level playing field is one where the costs of pollution are internalised, and hence in addition to a border adjustment for the production of Chinese steel, the ideal answer would be a shipping tax. Since such a global shipping tax is politically very difficult, there needs to be a border adjustment.

Calculating exactly how much carbon is emitted by each ship is difficult, expensive and would need careful verification. According to our general principle of better to be roughly right than precisely wrong, a shipping border charge should be based on the distance travelled (recorded in the ship's log) and then, say, a small number of general categories to get a pence per nautical mile. It should be part of normal customs declarations.

Similar considerations apply to aviation. Different aircraft have differing emissions. A fuel use declaration could be required, or a few categories with differential rates – short haul, long haul, and above and below, say, 25,000 feet (as greater height leads to a greater forcing effect of the emissions), and then apply a pence per mile. The bigger problem in aviation is what proportion of the international journeys are relevant for the country applying the border tax. It is easy for the airlines to pose this problem as insuperable and argue that only an international agreement will work, in the sure knowledge that this will be a long and tortuous process, and in the meantime carry on as usual. Charging passengers on departure or arrival may be an intermediary step.

What this all demonstrates is both that there are already many implicit carbon prices on transport, and that a carbon border adjustment is practical as long as the aim is to be roughly right. The carbon price that is set needs to be made universal in the economy, so that all transport faces the same carbon prices, and all other sectors of the economy face the same price such that the relative decarbonisation of the various sectors takes place in the lowest-cost way.

Put another way, the fact that all the fossil fuels currently have different implicit carbon taxes makes decarbonisation more costly than it needs to be, and hence makes the politics harder too. It is incredibly inefficient. Increasing fuel duty in the UK has been a fraught exercise so far, and the populist offers to lower it go in precisely the wrong direction. There is no case for coal to be less taxed than oil, and no case for providing subsidised red diesel to farmers, as noted in the previous chapter.

Vehicle standards, dieselgate and biofuels

The wonders of oil and its low cost make it very hard to make the shift to electricity. It is going to be a very tough nut to crack, and it is unlikely that a carbon tax on its own will do the job in the time frame required. It is also, as noted above, hard to generally tax the emissions from cars and other vehicles without taking account of geography. For these reasons there is a case for also applying regulatory standards to emissions from vehicles and to regulate and limit urban transport too. The localised presence of other externalities reinforces the complementary role of transport regulation.

The politics of raising taxes on motorists is typically poisonous. The cult of the car, the advertising and branding of the big car companies, and the promotors of motor sport and speed have captured much of the popular culture. As with raising the price of electricity, politicians have typically preferred disguised stealth taxes in the hope that motorists will not notice. The way to do this is to regulate: to impose vehicle standards for a host of things, including emissions. Limits are imposed on emissions of carbon and nitrogen and sulphur oxides, and car manufacturers are required to comply with these; with these limits, petrol and diesel cars can be prohibited.

Emissions standards, provided they are imposed on imports as well as domestic production (a border adjustment), do have the merit of being immediately effective. It is true that car companies have incentives to get around the rules, just as Formula One racing cars are constantly tweaked to the limits of rules and regulations.

The infamous 'dieselgate' scandal at Volkswagen is just one example of the more general attempts at evasion and avoidance. Yet even dieselgate shows that standards work. Volkswagen was found out, and all car companies have learned the lessons of such behaviours.

There is little doubt that fuel efficiency is greater because of these rising standards, and similarly there is no doubt that they have been responsible for a lot of innovation. There can also be little doubt that, in the absence of an adequate carbon tax, such regulatory approaches are going to be necessary to force through a sufficiently rapid switch to electric vehicles. Applied at the border too – and they usually are – they can be unilateral for larger markets like the EU and the US (but not perhaps for the UK, which, post-Brexit, will have to follow in order to maintain market access).

Some standards are less effective than others, and some are adverse. The most notorious and misguided have been the rules on biofuels. The idea behind biofuels is simple: replace fossil fuels with, for example, ethanol made by photosynthesis. The plants take the carbon out of the atmosphere, so when the ethanol is burnt the net result is zero emissions.

The reality is very different. Growing biofuels is not net zero; nor is turning them into useable fuels and delivering them. The crops may use fertilisers, require tractors and machinery (all diesel-driven), and the manufacturing process is energy-intensive. This is bad when it is sugar cane, corn (maize) and wheat.[12] It is terrible when it is palm oil, which has been a major source of the EU's so-called 'renewable' fuel.

It is understandable that farmers and their lobby groups love biofuels. It is a whole new market for them. It took at one stage nearly 40 per cent of US corn production, driving up demand and prices.[13] As a result, farmers make more money not only in the biofuels market but also from the higher prices in conventional food markets. Extra demand means higher prices.

The carbon calculation needs to take into account the uses the land would have been put to had it not been used for biofuels. In the case of palm oil, it is often cleared rainforest, which is about as bad as it can get. In the case of Brazil's sugar cane, it displaces other types of farming, including cattle. In the UK, it encourages the intensification of cereal production. Since the wheat is not going into the food chain, the chemicals used do not much matter. Having promoted biofuels, even the EU has had second thoughts.

The biofuels case illustrates a more general point about regulation. Every regulatory rule is in someone's interest. That interest will not only want to capture the economic rents from the implicit or explicit subsidies, it will also want to protect and grow those subsidies. Once a rule is set it is very hard to get rid of, unless there is some countervailing vested interest. Unlike carbon taxes, regulation is easy to capture, and it usually is captured.[14]

Urban regulation and planning

The other pollutants from transport are mainly local in their impacts. Nowhere is this more obvious than in towns and cities. Emissions from cars and lorries damage people's lungs and shorten lives. Pricing mechanisms can make some difference, and urban congestion charges have a part to play. They ration the supply of road space and can reduce traffic. By calibrating these charges according to engine size and vintage (and hence pollution), the more polluting ones can be priced off the road.

Congestion charges will, however, never be sufficient, and the regulatory and planning options have to come into play. Banning cars from certain areas and turning road space over to cycling and walking are increasingly policies of choice in developed countries. Restricting hours for goods vehicles and deliveries both rations road space and can concentrate emissions at times when fewer people are on the streets.

Such policies work best when there is an alternative mode of transport, and in particular public transport options. Urban buses, trams and underground trains give people choices, and when combined with high prices for vehicles entering city centres, as well as high parking charges, a substitution is possible. That public transport can be electric: directly, or indirectly via hydrogen.

Provide the public goods: innovation, R&D and the new infrastructures

Carbon taxes, augmented by regulation, make polluters pay, and that means those of us who drive the cars and expect our goods to be delivered to our door. They encourage us to use renewable and nuclear electricity. But we can do this only if the transport and electricity infrastructures are in place. There needs to be a charging network connected to the electricity system to deliver the power as demanded, and public transport options too.

Recall that infrastructures have important public goods elements. Private markets will not produce the optimal levels for the economy and the net zero objective. No private investor is going to pre-invest in charging networks ahead of demand. The networks are natural monopolies, with high fixed and sunk costs, and close to zero marginal costs, and for this reason they need regulatory protection.

A number of companies are developing charging networks, but as with mobile and broadband coverage, it is piecemeal and reflects particular interests. The oil companies want to hold on to their filling station sites. They have a lot already invested in them. They want to install their own charge points, with less regard for whether they are compatible with other companies' technology. As with the retail petrol brands before the star rating system, each wants to have a brand presence and promote it at the expense of the others. Shell is well advanced in this respect, having purchased First Utility, an electricity supplier, and rebranded it under the Shell name. BP is following suit with its charging system. This mirrors the fragmented approach in communications, and every car company has its own batteries.

To get the public goods out of these private investments, the system needs architecture, and someone in charge. This is where system operators come in, laying down rules for interoperability and standardisation. Networks become open, and public rather than private. It is not so much that there is one best option, but rather it is about making considerable gains by having one system. We all drive on one side of the road, but we are still plagued by different chargers for phones and laptops. Conventions create the maximum system interconnectivity, and in the case of electric vehicles this matters.

Who should have the duty to provide secure supplies of electricity for transport? For the electricity system, the security of supply obligation is still not properly defined, but at least the electricity system operator, National Grid, has a core responsibility. In the case of charging networks for cars and other vehicles, no one does, because there is no nationally agreed roll-out plan. There should be. This will have a considerable local element, and it makes sense to give this to regional system operators. As I set out in the Helm Review, these should be public, not private, and the required infrastructure should be auctioned to all the possible providers. Provision can be competitive. Standards should not, and nor should the system planning, or the duty to ensure supplies of both the network charging and the electricity. This will be further explored in the next chapter.

The other main area of public goods in transport relates to R&D and innovation. A transformation from the internal combustion engine to electric and hydrogen vehicles will require a lot of new technologies, many of which have not yet been fully developed. Private competitive markets will under-provide R&D.

Governments have provided R&D in transport for a considerable period, and this tends to be in transport research centres, laboratories and universities. There is nothing special about transport here: it applies to all R&D and it is the job of government to provide these public goods primarily with public monies.

Net gain and the offsets: reducing travel versus buying out your pollution

Airlines offer passengers the opportunity to offset their emissions. Cruise liners should do the same. So far it is all voluntary. You decide whether you want a clearer conscience. You could buy an electric car, and you could decide to stay at home and not take that cruise holiday. But most people probably will not, and, as we have seen, electric cars are not completely green, while staycations are not carbon-free either.

One of the reasons you may continue to drive to work is because work is organised in centralised offices and factories, and public transport is inadequate. To achieve net zero, we will not only have to avoid those overseas flights and cruises, but also avoid quite a lot of the other travel as well. One of the biggest obstacles is the absence of the most important network of all – fibre and broadband.

For a long time it has been recognised that the easiest solution to transport congestion and pollution is to reduce demand, and the easiest way of doing this is to turn the modern economy on its head, with work decentralised to people at home and in local offices, and then goods and services delivered to them. This is finally within our grasp, and within the period necessary to get to net zero. The Covid-19 lockdowns have given us a preview of what this could look like at scale, with large sections of the workforce making the transition to home-working at relatively short notice, albeit mostly on a temporary basis.

Imagine not having to endure the stress of the daily commute. Imagine if you did not have to pay the travel costs, and then inhale the polluted city air. Imagine the time saved, which could be spent with your family instead, and on your leisure interests. More time for family and friends, more time for exercise and more time outdoors. This would be a world which would be massively more efficient and massively more rewarding. You would be a lot better off. Local communities would thrive and the countryside would again be a thriving place.

The reason this is in the imagination rather than a reality (Covid-19 notwithstanding) is because businesses rely on physical contacts. The endless meetings have a point. They produce unanticipated cooperation. In addition, making workers turn up at a centralised location ensures they actually work. How otherwise does the manager know the workers are not doing other things when they should be producing for the company? On Facebook, Twitter and Instagram rather than concentrating on their work?

Fibre and broadband offer the chance to overcome some of these obstacles, and to avoid the travel costs, stress and time. Instead of buying an annual season ticket from, say, Winchester to London for £6,000, and spending a couple of hours each way getting to work and back, you could free up four additional hours a day, and be £6,000 better off and a lot more productive, free of the stress of the travel. If enough people did this, more roads might not be needed; the railways would be less congested and even HS2 might be redundant. Again, Covid-19 has given us an enforced glimpse of some of the possibilities.

The new ICT allows for constant connectivity, conference calls and video links. As explained in chapter 6, you can now have remote meetings where it is as if you are actually present. I no longer need to travel around the world for conferences. I do it by video instead.

This new world becomes possible only if fibre is provided and to everyone. It has to be a USO. But, as with charging networks for cars, it is anything but, with competing firms protecting their own interests and picking off the cherries and neglecting the harder-to-reach and less-wealthy customers. It is again a *system* problem and it needs a *system* operator to provide this essential public good network. Constant connectivity works only if everyone is constantly connected. Provide universal fibre, and a host of transport problems get solved. The first strategy to reduce your emissions, not to travel, would be so much easier to do. There is no need to offset. You can cut this out of your carbon diary.

But supposing you still do need to travel. What should you do? The net gain principle says that you should pay to offset your emissions, and over and above the expected detriment. How could this be effected for carbon? It has to be by sequestration. This can, as noted, be natural sequestration by trees and soils, or industrial CCS through storage wells and sinks.

None of this is straightforward, and the moral question about buying your way out of responsibility remains. That recognised, there is a practical reality to face up to: fossil fuels are going to be used in transport for some time to come, and at scale through to at least 2040. It looks pretty inevitable. The choice is: let these emissions happen and berate the polluters; or try to avoid them, recognising that if they do happen, compensation through offsetting will be required. The former might give a sense of moral superiority, but it results in the damage minus any offsetting investments. The latter at least ensures that something will happen to deal with the consequences. The greater the 'gain' bit in net gain, the more the consequences are ameliorated.

The difficult bits: aviation and shipping

Even with a carbon tax and the pricing of all the other pollutants, and with regulations and standards, there will remain some tough challenges to tackle emissions from transport. Of these, shipping and aviation look particularly problematic, and especially since they are largely international and not purely under national control. As noted, it is quite hard to be unilateral about either of them.

In both cases, the search for some international agreement on emissions and to impose technical regulations is being undertaken.[15] Where ships dock, they can be prevented from running their engines and polluting the local air. But on the high seas, no one has the power to stop the emissions. Even preventing oil spills is problematic. For airlines, the forcing effect of emitting higher in the atmosphere means that although the aviation share of emissions is a small proportion of the global total, its rapid growth means that net zero cannot be achieved without doing something about the carbon. Remember too that the flight itself – and the shipping – is only part of the total emissions. Airports are terrible places for carbon emissions, as are docks and ports.

In both cases, while there may be long-run technological solutions, like solar- and battery-powered planes and hydrogen-powered ships, the more efficient answer is to reduce the demand. One of the silver linings of the developing trade wars and tariff increases is that they may reduce shipping. Once carbon consumption is considered, and 3D printing and robotics undermine the advantages of cheaper labour in the Far East and especially China, reshoring of production is economically more attractive. This may, for non-carbon reasons, be the beginning of a slow but possibly even permanent decline in global trade, and with it shipping. Add in carbon border adjustments so that the playing field is properly levelled, and trade should fall a bit further. Also add in the worries about the extended supply chains and security of supplies during the Covid-19 lockdowns. There is nothing inevitable or indeed optimal about shifting ever-increasing shipments of physical goods around the planet.

There is a strong case for treating aviation and cruise holidays as luxuries and applying serious pollution charges directly to airline and shipping tickets.[16] If the passengers on cruise liners really understood the damage their holidays cause to the environment, they might think twice about taking this option. Perhaps they need a carbon warning, analogous to the health warnings on tobacco packaging.

The simple fact is that the continuing growth of conventional shipping and world trade, and of airports and aviation, is incompatible with limiting climate change. Short of a technological transformation, and quickly, if climate change is to be mitigated, shipping and air travel will have to diminish.

To see how the combination of fibre and communications technology and reducing transport demand works, consider the admirable case of Greta Thunberg sailing on a high-tech, low-carbon racing yacht to the UN conference in New York in September 2019. She decided not to fly. Good. She decided to go by a zero emissions boat. Good, but incredibly expensive for all but a small global elite. Why didn't she just stay at home and engage with the UN conference delegates by video link? Showcase what modern communications technologies can do? That would be a genuinely persuasive story to tell, and no doubt communications companies would fall over themselves to show how good the live screening and engagement can be. This may be for the next great global COP gatherings, and then she might add to her already impressive contributions by helping to create a new wave of vested interests, dedicated to undermining the need to travel. We come back to your carbon diary and the central question about travel consumption: why do we need to travel so much?

10

THE ELECTRIC FUTURE

Not only is electricity a big emitter, but it is one whose importance to decarbonisation goes much further. Electricity is what powers the digital economy, and it will be what the digitalisation of almost everything else depends on, including transport, agriculture, manufacturing and households. It is also the potential feedstock for hydrogen. Unless electricity is decarbonised, electric cars and the electricity-powered robots will simply substitute one form of polluting energy for another.

Electricity has already come a long way. In the 1980s, almost 80 per cent of UK electricity was generated from coal, with the rest mostly nuclear. It was a similar story in Germany, the US and, until very recently, China too. Coal was king, and for much of the global electricity industry it still is. The growth in coal demand, especially in China, transformed the concentration of carbon in the atmosphere from worrying to disastrous. It is a primary driver of the last 30 wasted years. Coal-generated electricity is what has really damaged the climate, putting even oil into second place. No amount of natural sequestration can keep up.

Getting out of coal

The starting point is therefore to get out of coal. Otherwise, there is really no chance of limiting global warming to 2°C. It is not just about reducing the share of coal in the electricity generation mix. It is about reducing its absolute burn. As noted in chapter 1, with the world economy growing at around 3–4 per cent GDP per annum, the IEA's forecast of its share falling from 40 per cent to 25 per cent of world primary energy by 2040 is actually an *increase* in the total coal-burn.

There are several ways to get out of coal in electricity generation. They broadly fall into two categories: switch from coal to gas and then *later* to renewables; or switch straight from coal to renewables and nuclear. The two camps might be called the pragmatists and the fundamentalists. The pragmatists point out that for new power stations, gas is cheaper than coal, and so as the old coal stations close, they can be replaced with gas at little extra cost. Gas has around half the emissions of coal, and it tends to be much more flexible. Gas stations can ramp up and down to offset the intermittency of renewables, and cope more effectively with the big swings in renewables production.

The problem the fundamentalists see is that, in a net zero world, there is room for only a tiny amount of gas, and that should be offset by compulsory natural sequestration and CCS. They worry that, once built, the gas power stations will be used and hard to get off the system later on. The temptation to use the gas stations will, they worry, be strong. So better not to let temptation stand in the way. They would get on and build lots more solar, wind and nuclear now, and sort out the consequences for the system. One way to make sure this happens is to require that CCS is fixed to any new gas stations, knowing that this will undermine the economics of such investments.

How to get out of gas

In a net zero world, the gas will have to go too, but crucially not yet. The fast-track closure of coal, driven by ramping up the emissions standards and regulation, and increasing the carbon price, is the immediate prize, but it will leave a very big hole. The UK is far advanced with its coal closure programme, and beginning to plan for the hole in baseload supplies that this will result in over the next five years. With new nuclear delayed, with old nuclear closing, and with little large-scale gas plant construction, the UK is one of the countries closest to the fundamentalists' position. Wind, especially offshore wind, will have to plug much of the gap with much more to come, as heralded in the December 2020 Ten Point Plan and Energy White Paper.[1] It remains to be seen whether the system will cope.

Elsewhere it is a different story. As explained in chapter 3, Germany has planned staying with coal until 2038, and getting out of gas and nuclear. The US is switching from coal to gas because it has abundant and cheap shale gas. China is making some substitution from coal to gas (at the margin only) and building new nuclear, although it is still building lots of coal at home, and increasingly encouraging the building of new coal-fired power stations abroad.

The reason why pragmatists like to pace the transition, and give gas a significant role, is that they worry about the costs, and whether the voters will stomach the rising prices. Coal to gas makes little difference to the electricity price. Coal and (old) nuclear to wind and solar makes a big difference. More gas in the short run may also ameliorate the system problems, and hence both avoid the threat of power cuts and encourage investment in the extra capacity to deal with the intermittency (which has so far tended to be small-scale gas and even diesel engines).

The coal-to-gas switch does not take much of a push to make it happen. Given the relatively close cost competitiveness between the fuels, all it takes is a nudge in the right direction. A carbon price is probably enough to do this. In the UK, it was the Carbon Floor Price plus the EU ETS which tipped the balance and accelerated the coal closures ahead of the 2025 cut-off point, reinforced by regulation.[2]

To get out of gas subsequently, two steps are necessary. First, on the polluter-pays principle, the price of carbon, having first done for coal, should as it ramps up help to do for gas. Second, regulation through the CCS requirement may well finish the job.

The low-carbon options post-coal and gas

With coal and then eventually gas on their way out, what to put in their places? Step forward renewables and nuclear. The vested interests and their lobby groups would have us believe that these are all cost-competitive already. This is the 'miracle solution' we met earlier in chapter 2. The renewables lobby claims not only that wind and solar have already reached 'grid parity', but that they are both much cheaper than nuclear. The nuclear side argues that this is not based on the true costs of renewables and, once these are taken into account, nuclear wins. Others argue that we need all the technologies, on the grounds that every bit helps.

Not surprisingly, comparing the costs of different technologies is a lot less obvious once we move from lobbying to facts. Remember, the renewables lobby is not campaigning to remove subsidies, and its members conveniently tilt the comparison in its favour by leaving out lots of the relevant costs. On the other hand, the gas lobby neglects to factor in a full carbon price. When it comes to nuclear, the scale and time horizons make any comparisons very difficult.

All of these lobbies neglect to take account of what is likely to happen in the fossil fuel world.[3] As and when decarbonisation happens, it should reduce the demand for oil, just as the cost of oil production keeps falling with technical progress. The renewables and nuclear will, as noted, then face competition from ever-*lower* fossil fuel prices. That makes the era of 'subsidy-free' renewables and nuclear a more-distant prospect.

What renewables and nuclear need is a lot of technical progress, and not just in their core generation units. They need a revolution in managing intermittency and in the way networks are operated. The good news is that this is coming. The bad news is that the system architecture has yet to adapt to incentivise these developments.

Equivalent firm power auctions and system operators

In the Helm Review, I set out a system architecture to take the existing markets from the coal and gas world to one based on renewables and possibly nuclear.[4] The aim is to normalise renewables so that the electricity market is organised around them, treating them as core, and the fossil fuels as the add-ons, with nuclear, as ever, a special case.

There are two key features to a new renewables-driven electricity market: the switch to a capacity world; and the establishment of system operators at the national and local levels.

For the last 100 years, the electricity price has been determined by the price of fossil fuels, and that in turn has been driven by the cost of oil, coal and now gas. Our electricity bills have been derived from the wholesale price of electricity, and this is mostly determined by the marginal cost of the last coal and now gas station on the system to meet the demand at each point in time. It has been an energy market based on marginal costs.

The renewables world is one that is largely zero marginal cost. It does not cost anything to generate an extra bit of solar or wind electricity. All the costs are in the solar panel or the wind turbine – the capital assets. Renewables are mainly about capacity, not energy (the exceptions are biofuels and biomass). The wind farm developer sinks a lot of capital into building the wind farm, and then sits back and watches the turbines go round. There is a bit of maintenance, but it is small compared with the initial fixed and sunk costs and, in any event, it is not tied to the marginal unit of electricity generated.

This has profound consequences. If the marginal cost is zero, so should the wholesale price be zero, with 100 per cent renewables meeting total demand.[5] This turns the energy-only wholesale market, driven by fossil fuels, on its head. Furthermore, consider how absurd it is that the price of electricity in a context of, say, 95 per cent renewables output is determined by the market price of gas just because gas produces the last marginal 5 per cent of the electricity.

Zero marginal cost wind and solar drive the energy price down towards zero. This creates a problem not only for the existing gas power stations (and it is the reason why few want to build any more), but also for anyone building renewables too. How do they get paid? The answer is that there needs to be a market in *capacity*, and you and I need to pay the cost of the capacity rather than the cost of the energy. Energy is free, but capacity is expensive. When you turn on the light, what you want is a system in place which will deliver that light, in the same way as when you plug into your broadband network, you want instant access to the internet. The new economics of electricity is much more like broadband and fibre: lots of fixed costs, and zero marginal costs up to the point that the system is congested.

What you really want is the security that not only when you switch on the light it works, but that there is enough system capacity so that it *always* works. You do not ever want to be rationed off the electricity system by power cuts which can be massively disruptive. The costs of that disruption are mounting as more and more of the economy is electrified and digitalised.

In August 2019, Britain had an unpleasant reminder of these consequences. A large wind farm and a conventional power station dropped off the system within less than a minute of each other, triggering a failure of the National Grid and local networks, and resulting in widespread disruption across England and Wales. While system frequency (essential for the stability of the system) was restored within a matter of minutes, the impacts were evident for several hours – particularly for the rail network which was still feeling the effects the following day.[6]

What worked for the coal and nuclear age, with the wholesale markets reflecting fossil fuel marginal costs, will not work for the new world of zero marginal cost renewables (nuclear is zero marginal cost too). What is needed is an inversion of the market, from wholesale marginal costs to fixed capacity costs, and therefore from a wholesale to a capacity market.

The Helm Review set out how this could work. There would be a single central capacity market, based on firm power. The bidders would offer what they can guarantee to supply when called upon. They would be paid the clearing price in the auction for their capacity offered to the system. It is a fixed income, against a fixed cost, and therefore properly backs the financing of the new investments.

The way this integrates the intermittent renewables is critical to the market design. Renewables bidders would offer equivalent firm power (EFP). This would be less than 100 per cent firm power because they cannot always guarantee to deliver. The wind might not blow, or it might blow too much. The value of the renewables capacity would be adjusted to take account of their smaller contribution to the system. An offshore wind farm adds security, but not as much as a nuclear power station, which is always on. How much security the wind farm adds depends not only on its own characteristics and how often the wind blows, but also on what else is on the system. This less than 100 per cent firm power is de-rated, so that it is reduced to the number which is equivalent to firm power. A wind farm might, for example, be 40 per cent, depending on how many other wind farms are on the system at the same time.

Note a very important incentive the EFP auctions create. Consider the position of a wind farm owner, and suppose it is de-rated to 40 per cent EFP. The obvious question is: how can it get above 40 per cent, towards 100 per cent, and make more money? The answer is that it would need to find a way of turning its intermittent supplies into firmer ones. How to do this? Find its own matching back-up. Look to the active demand side, and find someone who is prepared to have their supply interrupted when the wind does not blow. Build its turbines, or perhaps solar panels, next to someone who can flex their demand. It might be the vertical indoor farms using LED lighting. It might be the producer of hydrogen, using surplus wind. It could be someone charging car batteries.

The wind farm owner would have a powerful incentive to identify such flexible demand to match its intermittent supplies. In practice, a secondary market would develop (there is an imperfect one already), and this could be entirely driven by those market incentives derived from the EFP auctions.

It is all a bit complicated. But then our new world is. You do not need to understand how your internet banking actually works and your transactions are cleared. What you want to know is that it does work, that you can get your money when you need it, and that it is secure and efficient, making it as cheap as possible for you. Similarly, a renewables energy market performs a very sophisticated and technical task to make sure the system always has enough power to meet your (and everyone else's) demand, plus a margin of security.

The next question is: who is in charge of the EFP auctions? Who decides what is auctioned? Who decides the system characteristics, including the supporting networks? Who makes it work? Like the banking example, there needs to be a clearing mechanism, and in electricity it has to be instantaneous. That, in turn, requires a lot of planning and preparation.

The answer to all these questions is that there needs to be a system architect who decides what capacity is needed, plans out possible scenarios, and makes this happen through the EFP auctions. This is the system operator. There is nothing unique about this function. It is common and implicit in all networks and infrastructures. River catchments need coordination and planning. So do the railways, the banking system, and the broadband and fibre networks. All would benefit from an explicit system operator.

A big obstacle is the incumbent network companies which have a vested interest in keeping this function to themselves. In controlling the system, they can make it work best for themselves rather than the wider public interest. Their control entrenches monopoly power. For this reason, to maximise the speed of transition to the decarbonised world, and to minimise the costs, the system operator needs to be independent and separate from the incumbents. It needs to be separated from the National Grid (just as it should be separated from the water companies, Network Rail and BT's Openreach in the other networks). It should be in the public and *not* the private domain since it is charged with deciding what is in the public's interest, and it must be a neutral arbiter in the EFP auctions.

There is an extra twist to this new electricity world. Much of the renewables, and most of the new demands from transport, houses and factories are local rather than national. The new world is full of relatively small-scale solar panels and wind turbines. It is not a world of large coal and gas power stations (although it may still contain very big nuclear power stations). It is disaggregated and decentralised, embedded in local rather than national networks.

It is also at the local level that the opportunities for technical change are perhaps the greatest. Smart meters will provide smart data at the household level, opening up scope to use the data to better optimise the local systems, and offering the possibility of active demand responses. The smart systems may help with the new car-charging requirements. Battery storage can offset the intermittency of local renewables generation, and the network itself may become much more fluid. Instead of automatically adding more cables, perhaps a wind turbine embedded in the local network, plus storage and some active demand response, might be better?[7]

Taken together, it is immediately obvious that the current local network owners (distribution network operators) are unlikely to be unbiased about these choices, especially if they have a deep vested interest in protecting and enhancing their assets. It is therefore even more important to have independent system operators at the local level – what the Helm Review calls the regional system operators. Each regional system operator auctions EFP at the local level, inviting bids from the demand, supply and storage sides. And it goes even further, opening up the network enhancement proposals from the network operators to alternative bids for generation and storage and the demand side too.

This opening-up at the local level is essential for the transition to the decarbonised electricity systems, and it is unsurprisingly fiercely resisted by the network owners. Put simply, it is about having someone in charge of the system, making sure that the public interest in security of supply is met (and therefore that there is enough firm power capacity available), and that decarbonising is taken care of. The present arrangements, which leave these critical system issues in the hands of private monopolies, are unsatisfactory, and it is not surprising that the results are higher prices and less security. Net zero cannot wait for the network owners to fall into line.

This is the new architecture required to transform electricity, and therefore provide the inputs to the rest of the economy as it too digitalises. It is not a question of evolving into this EFP and system operator model. It will not happen by evolution because the vested interests will defend their economic rents. Renewables will want to hang on to their special subsidies and contracts, added on a case-by-case basis. The network owners will want to protect their RABs. A deal-by-deal world is one that maximises the scope for capture. An EFP auction makes people bid real money and face real competition. It is always a market open to new ideas and new entrants. That is why some incumbents hate the idea. But even with EFP auctions and system operators, can the electricity industry deliver on its foundational role in decarbonising? Can it deliver with wind farms, solar panels and nuclear power stations?

Electric pollution and the carbon price

Markets typically deliver cost-efficient outcomes on the basis of the prices and costs that the bidders face. If those prices and costs are wrong, markets simply reinforce the distortions created by them. That indeed is what happens in respect of carbon. The carbon price does not include the full cost of the carbon pollution caused. Not only that, coal is not paying the full price for the other emissions that damage air quality. It is not paying for the health consequences to the miners and all those exposed to coal through the supply chain. Gas is not paying for the methane leakage, and renewables are not paying the full carbon costs of their inputs in manufacturing them, their network and back-up costs, or the visual and broader environmental costs they can also sometimes present. Distorted costs lead to distorted bids, which lead to suboptimal outcomes. The EFP auctions and the system operators are necessary for the electricity transition to low carbon, but they are nowhere near sufficient.

Over and above all the other distortions, the biggest and most central is the price of carbon. How would a carbon price in the electricity market work? Imagine if all the bidders in the EFP auctions faced a common carbon price at the correct level to reflect the carbon pollution generally, in line with the carbon budgets on the trajectory to net zero targets in 2050. Now the bids would be very different. A gas power station would have to bid its cost of gas, plus its carbon per unit, plus its capacity cost. In contrast, the solar farm would have no equivalent to the price of gas (or oil or coal), and would have no carbon price to pay. It would have only its capacity cost. It would be a new world. You and I would, of course, pay this carbon price in our bills, but only if we bought electricity generated from fossil fuels.

There would be a number of intermediate cases, like biomass wood pellets burnt in converted coal power stations. The inputs should come with a carbon credit for the temporary sequestration by the crop, but the carbon price for the energy used to pelletise the wood and dry the pellets, the diesel to truck the pellets to the coast, the shipping emissions, and the energy required to keep the pellets cool to prevent combustion while in storage would all be added. So too should the emissions from burning the pellets.

That is not what happens yet. There are lots of market distortions, and the Helm Review sets out how to deal with some of the second-best problems. Suppose there is no carbon price. What is the alternative? The answer is to still have an EFP auction, but to ask all bidders to include the associated carbon emissions that their bids entail. What the national and the regional system operators now do is set the bids alongside the carbon budget produced by the CCC. These set the envelope of allowed emissions on the path to net zero in 2050. The bids are then adjusted so the total capacity required by the system can be provided within the overall carbon budget. It will almost certainly cost more, but it will be much better than the status quo, which is driven by the wholesale market, without an adequate carbon price. It might sound complicated, but determining future electricity generating capacity is crucial to achieving net zero.

The EFP auctions and the system operators get us a long way towards normalising renewables, and if a carbon price is added, a solution is in the making. All very exciting, but still not enough, and for two reasons. First, the networks still need attention so that they are designed in line with decarbonisation. Second, we need new-generation technologies since current wind and solar, and even current nuclear, probably cannot do the job on their own. Very careful attention needs to be paid to the public goods – the infrastructure and the R&D.

Electric infrastructure

The electricity system is a *system*. This may be obvious, but it largely passed by the architects of the privatisation, liberalisation and competition agenda. They thought that the old state-owned and state-planned great electric utilities, like the CEGB, EDF and RWE, were dinosaurs of a past age, hopelessly inefficient, and in need of breaking up. In their brave new world, there would be no monopoly to plan and to charge from. Instead, the system would be gradually unbundled, with new entrants building new power stations, challenging the old, and with the infrastructure networks as the servants of the generators and suppliers. Anyone could switch to any supplier. Those on the right thought that it would become just like any other commodity market.

There were obvious gains at a time before there were serious concerns about climate change and when there was a legacy of many power stations. Recall that, back in the great postwar economic boom time, from 1945 through to 1979, the going assumption for the UK was that 3 per cent economic growth meant a 7 per cent increase in electricity demand, and the CEGB built coal and nuclear power stations to stay ahead of the curve.[8] When the UK economy went into free fall in 1980, and as it deindustrialised and entered a period of much lower growth, some of those power stations, epitomised by Drax, then the largest coal-fired power station in the UK at almost 4GW, were surplus to requirements. Privatisation and competition could therefore concentrate on what it proved good at – sweating the existing assets and driving down operating costs.

By the 2010s, although the privatisation rhetoric was maintained, the return to a system approach was well under way. Electricity Market Reform (EMR) performed the *volte-face*: now almost all new UK power stations would come with a government-backed contract, and the customer would be the government, not you and me.[9] With EMR, generation came back under (indirect) state control. We could switch supplier, but largely as a way of paying a smaller share of the total system costs.

In any event, the networks themselves remained monopolies and subject to regulation. The network regulators tried to unbundle the remaining assets. Meters were spun off, the harbinger of the disaster of getting smart meters installed by suppliers rather than (as everywhere else in Europe) by distributors. New network connections to offshore wind farms were contracted by third parties, not the Grid. The networks were prised apart, just when the system needed more, not less, coordination. This makes the future of system operators all the more important.

What sort of networks are needed for a world in which renewables and nuclear are normalised? The answer depends on how fast and at what scale the networks decentralise, and how fast storage and batteries develop. In a world with active demand and lots of storage, it is not clear exactly what the centralised networks are for.

That, however, is a long way off. In the meantime, the network infrastructure needs to be planned for some very big incremental increases in demand. First up is the electric vehicle charging network. Rather than a free-for-all as described in chapter 9, there is a strong case for a planned charging infrastructure, for placing this with the electricity distribution companies, and for making the investment ahead of demand. The national system will also need to provide for the other big increments for high-speed trains and fast-charging centres for car batteries. If we are serious about climate change, we need to get on with it.

This cannot be left to the regulator to determine in negotiations with the private owners. The former has an inbuilt focus on driving down prices, and in the process squeezing costs. The latter have an interest in their RAB and the returns on this as a proxy for investing in index-linked government bonds. These two interests are not the same as the public one, and neither has an obligation to make sure net zero happens.

It is for this reason that the Helm Review proposes that Ofgem and the equivalent economic regulators in other sectors are replaced by the national and regional system operators. It should be the system operators that determine what is required to meet net zero, and they should use a combination of the EFP auctions to provide secure capacity and open up the network investments to competitive bids from the demand side, and storage and local renewable generation options, again through auctions. It is this model which best blends system control and competitive delivery, not the liberalised model pursued since privatisation. Private companies bid to build and provide what the public interest demands, determined by the system operators.

R&D and next-generation renewables

However well the electricity market is designed, including the EFP auctions, system operators and a carbon price, the fact remains that the existing technologies are probably not sufficient to get us to net zero on a consumption basis, and can get us there on a production basis only if we shrink back domestic energy-intensive industries even further. To crack global warming, new technologies in electricity are essential, as well as in agriculture and transport. In electricity this means new smart and AI technologies to optimise the demand side and integrate it with supply and storage, better storage options and breakthroughs in battery technologies and, most of all, new ways of generating low-carbon electricity.

Although there are some very promising lines of research, it is in the nature of technological change that it will surprise. The coming of modern communications technology has been a surprise in my lifetime, and a transformational one which is yet to fully play itself out. As with other major general-purpose technologies, it is not only a revolution in itself, but also revolutionises everything it comes into contact with, often over many decades. Electricity was like that, as was the internal combustion engine. Both took a century to have their full impacts. Digitalisation is likely to be the same.

Surprises should not stop us channelling R&D funding into those areas which look promising. Choices about spending have to be made. Closer to the carbon problem, we at least know what the questions are to which we need answers. They are all about how to generate lots of low-carbon electricity, to store it effectively, and to better manage the demand side.

Storage is all about batteries and other vectors for heat. Among the generation options, solar has more to offer than wind. Wind turbines are well developed and little more can be added other than incremental advances in materials, scale and logistics management. Floating platforms add a new dimension.[10]

Solar is potentially infinite in supply, although existing solar panels are primitive and do not use much of the light spectrum. Our ability to turn light into electricity is currently limited mainly to the panels. Opening up more of the light spectrum, new materials and solar film might be game-changing. All look feasible and are certainly worth a punt.

Any serious decarbonisation strategy has R&D at its heart. It is cheap compared with mass deployment of existing solar, and especially wind. Why spend, say, £30 billion on offshore wind and fail to find £1 billion for electricity-related R&D? I made this point in *The Carbon Crunch* and was roundly criticised for what is an obvious observation, on the grounds that switching even a very small proportion of the subsidies to R&D looked like being 'against renewables'. For those who claim that existing technologies will do the job (and the vested interests clearly have this in mind), there needs to be an honesty about the scale of the costs, the scale of the system requirements, and the prices that customers (and voters) will have to pay.

Like any technology policy, R&D is vulnerable to capture by the vested interests. Should we spend more on solar, wind or nuclear, for example? A clear separation between politics and R&D choices reduces capture through research councils and other familiar institutions. But these too can be captured, so when it comes to research budgets it is important to lean into the wind against conventional thinking, and especially conventional academic interests and group-think.

There is a high-level choice about areas and collaborations. Solar, nuclear and new materials (like graphene) will all figure in an overall R&D budgetary framework. Within each there needs to be lots of freedom to pursue the unlikely, the odd and the counter-intuitive lines of research.[11] Cooperation, new laboratories and a mapping of the very international focus of much research are all part of an effective R&D strategy. It requires university budgets, laboratory budgets, European cooperative ventures, and global initiatives. All are cheap relative to simply subsidising more and more of the same stuff we already have.

Net gain and our consumption

Recognising that existing technologies cannot get us to net zero, and that the target is only 30 years away, means accepting that some fossil fuels are inevitably still going to be burnt to generate electricity in 2050, and that there will need to be offsetting compensation. Avoiding emissions and, where this is not possible, mitigating them, comes first; only when the residual is left is offsetting sensible. Almost all the discussion around electricity is about CCS as the offset, but, as discussed, natural sequestration is likely to be the cheapest option with the greatest additional benefits.

Designing electricity systems, running R&D programmes and overseeing offsets through CCS and natural sequestration are all jobs for government and its agencies. You and I cannot do much about these. Does it therefore mean that we have nothing personally to contribute on the electricity side? That we can only offset to assuage our consciences? No, because all this electricity is ultimately generated for us. We are the demand side of this supply-side equation.

Much ink has been spilt in trying to claim that the answer is energy efficiency and demand restrictions.[12] If only we could use less electricity, we would save money and help to reduce emissions. But before we get carried away with energy efficiency, there are several dimensions worth reflecting on. The first is the overall point that, provided the electricity is zero carbon, there is no reason to use less of it. The sun comes up every day and its solar energy is free. Furthermore, we do not actually consume energy. We can only change it into another form. The aim should be to reduce our use of fossil-fuel-generated electricity, not electricity *per se*.

The second reflection is that energy efficiency can often be less than it seems, and is not always the win–win in cost terms it is presented as. Making a house more energy-efficient involves investments, and these have costs, take time and bring a lot of hassle.

Third, recall too that energy efficiency measures in existing buildings can be bad, and sometimes very bad, for your health.[13] Sealing your house up traps all the indoor pollutants, from the chemicals that go down the sink and toilet to the oven, the boiler, the carpets, and so on, and contributes to rising cases of asthma and other lung problems. During the Covid-19 lockdowns, households and schools were advised to open windows to improve ventilation and thereby reduce the risk of transmission. Retrofitting to existing houses tends to focus on reducing energy use, without taking into account these side effects. The energy efficiency lobby likes to paint a picture of state-of-the-art housing, the sort that wealthy people can afford. It neglects to focus on the run-of-the-mill housing. It is indeed remarkable that even in the midst of a UK house-building boom, almost all of them are not remotely close to net zero.

Finally, remember the Jevons Paradox discussed in chapter 6 (see endnote 8). Suppose the energy efficiency measures reduce your energy costs. You would have more money to spend on energy. That indeed is what has happened over the last century: as energy efficiency has improved, so the demand for energy has gone up rather than down. Think you might already have saturated your demand? Think again. Think about air conditioning – especially if the climate hots up – and the plethora of new communications devices you might soon be using. No doubt there will be many more demands for energy that are yet to be invented.

You should not waste energy, but you should take note of the four points above. Work out how much of your demand is for fossil-fuel-based electricity, and try to reduce this. You might simply decide you are going to buy only renewable electricity and biogas (although this is more complex and carbon-polluting than it seems, and not all supply companies offering renewable energy are really doing so). Make sure you understand the full costs, including the carbon costs of insulation and other measures. Make sure you have lots of air circulation between your house and the outside.

You can do simple things too. Wear a pullover and don't sit around in shirts in the winter evenings. Try turning down the thermostat. Try a warm shower rather than a hot one. All are probably going to improve your health at the same time – classic no regrets steps.

Once you have sorted out the insulation, you can move on to the electricity generation itself. You could generate your own with solar panels. You could install heat pumps. You could work with your community to set up a local generation facility, including local energy-from-waste. Even if the solar panels do not make much electricity, they could heat your water.

Then there is charging your electric car. You should do this at night and, where possible, using the solar power you saved during the day. This in turn will mean you need a house battery rather than a conventional gas boiler.

All of these are small steps. They will not make a lot of difference because your demand is too small to count. But a lot of small demands add up. This is a 'collective good' problem which requires individual cooperation to get off the ground. The bigger stuff, including the system architecture and the carbon price, involves your vote. As with your individual energy efficiency measures, your vote makes no practical difference to the outcome. But *all* of our votes do. Finally, there is your voice. Big environmental changes come from lots of small conversations. Have them.

CONCLUSIONS

A NO REGRETS PLAN

When the historians look back in 2050 and judge whether we fulfilled our duties to the generations to come, will they say we did our bit, and tackled climate change before it got out of control, or will they condemn us for our selfishness? Will they say we mended our ways, put ourselves on a sustainable growth path and took pollution seriously, or that we looked the other way, to our own narrow, immediate interests and carried on with our unsustainable lifestyles?

Over the next 30 years, the climate change challenge will be won or lost. We can have a green and prosperous land, without more net carbon consumption. Or we can have a hot and dirty land, and the very real prospect of a *really* hot and *very* dirty land, changing life on this planet as we have come to know it.

It is a choice, and it needs a plan, the plan set out here. The choice and the plan are both urgent. There is not enough time to allow a plan to evolve from one political gimmick to another. A plan is more than the summation of a series of eye-catching and vote-winning announcements, and the choice for us to make now is about how we want to live, and in particular what we choose to consume.

The building blocks of a plan that might actually work

The plan brings together into a coherent whole the various dimensions that have been set out so far. It starts with sustainable economic growth, not the GDP fetish which beguiles us. Climate change and biodiversity loss are so systemic that they demand an overarching economic system to address them. Critics of the current model of capitalism are right to point out that it is not sustainable, but they are wrong to rule out the 'growth' bit.

A plan for sustainable economic growth starts with consumption, and views production as the means to our consumption. The plan needs to create an economy within which our choices are channelled into low-carbon ones and, in the short-to-medium term, this means that we, as the ultimate polluters, should face the consequences of our choices. The way to do this is to make sure the costs of our pollution confront us every day. This requires a price of carbon, and it needs to cover imports as well as domestic production.

The carbon price brings the market into full play in the climate change challenge. It avoids the problems of government trying to predict winners, and it blunts the influence of lobbyists. The market is a process of discovery, and there is a lot to discover. It starts with the assumption of pervasive uncertainty about the opportunities and costs of the various options and, crucially, it seeks out the lowest-cost options. By contrast, the State has to know all the information to pick its winners, and it is wide open to capture by the various vested interests. That is why governments blunder from one initiative to another, and why losers pick governments.

The carbon price at the border is the way to make unilateralism work. It can lead us to ending our contribution to climate change, and it can encourage others to follow suit. It is vastly superior to wasting another 30 years trying to make Paris, Glasgow and its successors work out. They will almost certainly fail. Paris is already failing.

A carbon price is necessary but not sufficient. The role of the State comes into its own with the infrastructures, and in particular in planning and ensuring the delivery of low-carbon networks for communications, energy and transport. The State also determines much of the way land is used since it subsidises agriculture, which covers some 70 per cent of the land in the UK. Natural carbon sequestration has to play a central role as we will never get to zero carbon. It is *net* zero, not *gross* zero, that we need to achieve.

The State plays a further role in providing support for R&D, and technology is the ultimate source of economic growth. Whatever the lobbyists say in defence of current renewables, they are not enough. Indeed, they are not remotely enough to get to net zero carbon consumption. There needs to be a technological revolution, and the good news is that the science tells us that this is indeed not only possible but likely. To get there, governments need to step in to fund research – big time. It is the State that led the research which gave us the internet and mobile technology; it is the State that got us into space; and it is the State that developed nuclear energy.

With the market driven by the price of carbon, and the infrastructures and R&D in place, the scene is set for filling in the details of the plan. Since the current economy is soaked in carbon, all sectors face radical change. They always do when there are major technological innovations. They are all now digitalising.

Three sectors stand out: agriculture, transport and energy. The way the land is used and, in particular, its ability to sequestrate carbon, is staggeringly inefficient. In the UK, it is about as carbon-unfriendly as it gets: intensively chemicalised; losing soil carbon; and producing stuff that is mostly uncompetitive. As with much of global agriculture, it lives in the main on its subsidies. Transport is organised around the internal combustion and jet engines, and it needs to electrify, either directly through electric vehicles, or through hydrogen made from electricity.

It is not just transport but all of the economy that will have to electrify. The good news is that this will have to happen anyway because of digitalisation. Almost everything digital is electric. The even better news is that decarbonising electricity is probably the easiest bit of the transition to net zero. It will be expensive (indeed, it already is), and it will hit our standard of living, but at least it can be done.

These are the building blocks of a plan that could work. At its heart is the identification of the role of the State, the role of markets, and the role of you and me. There are political and moral choices to be made.

Time, as they say, will tell whether we make the right choices, learn the lessons of the last 30 wasted years, and get on with it. Suppose we fail. Would we, with the benefit of hindsight, regret not acting sooner, not paying a proper carbon price, not providing the public goods and not enforcing compensation for the environmental damage we do?

Suppose others do not follow our lead. Suppose China, India, Africa, and lots of Southeast Asian, Middle Eastern and Latin American countries carry on with the growth of their emissions. Suppose, wedded to the GDP model, they burn more coal, stick to oil and gas, and continue to wreak havoc on the natural world's ability to sequestrate carbon. Suppose the Amazon and the Mekong rainforests continue to be cleared and polluted, along with those of Malaysia and Borneo. Suppose, as a result, carbon goes beyond 500 ppm.

It has to be recognised that this is all a distinct possibility, even a likelihood. Almost nothing has been achieved in the last 30 wasted years. Why should the future be any different? Even if you care passionately about climate change, you have to admit not only that this nasty scenario is possible, but also that many of your fellow citizens think it will come to pass.

There needs to be an answer to the obvious question: why bother?

If we are all going to fry anyway, why not carry on the fossil fuel party and enjoy it while it lasts? It is no good simply avoiding this question: it has to be tackled head-on.

Fortunately, it has an answer, even if only a partial one. It requires us to think carefully about the 'no regrets' things we can do anyway, and to stop wasting money on things that make climate change worse.

Recall that not to price pollution is to be inefficient. Recall that not to provide the public goods is inefficient too. So is not paying compensation for the damage. An efficient economy incorporates all the costs into prices, and it relies on public infrastructures and public research funding.

Imagine what Britain could look like in 2050 if we did all these no regrets things between now and then. Think of all the benefits, in addition to us making no further contribution to climate change. The air would be a lot cleaner, and so would the rivers and lakes. There would be little or no diesel or coal poisoning the air and our lungs, not just with their carbon emissions but also all the other toxic particulate matter. The end of gas boilers would take out quite a lot of household air pollution. There would be a lot less sulphur damaging the trees. The pollution running off the intensively farmed agricultural land would be lower. Our water bills would be lower since we would not be paying for the clean-up.

We would also no longer be wasting our money on climate policies which may make things worse, such as subsidising energy crops for anaerobic digesters, subsidising the burning of imported wood pellets, and paying for biofuels made from palm oil. We would stop importing beef from the cleared Amazon ranches, and giving our industrial competitors an unfair advantage by importing carbon-intensive products.

The countryside would be transformed. In place of the chronically inefficient, intensive agriculture, nature would have more room to thrive, and we would thrive too, with better access to nature, more scope for exercise and improved mental health. The plastics that pollute our beaches and oceans would be replaced with biodegradable packaging.

We would have fibre and not need to travel so much, and especially do a lot less commuting. The ICT revolution would finally come home, literally. Inner cities would be better places to live and work in. We would have carbon-sequestrating trees along all the streets, green walls, and parks for the children to play in.

All of this would be encouraged by the technologies that we would have in 2050. Food production would be transformed by digitalisation, robotics and AI; much more plant production would be indoors, without pests and therefore pesticides. Insects would provide concentrated protein. Genetics, and especially gene editing, would take the place of many drugs and chemicals and increase productivity, and digitalisation would limit food waste.

The infrastructures would be very different. They would all be smart, and able to benefit from the enormous efficiencies to come from big data. Networks are at heart coordination problems, and our manual and primitive approaches would be greatly improved. Autonomous cars, powered by low-carbon electricity, will open up transport to many more people, including older people no longer able to drive.

This would be the age of electricity and decentralised systems, bringing in active demand management and storage, and integrating vehicle battery storage into the management of the electricity system as a whole. Your fridge and washing machine would be smartly integrated into this electricity system, all automatically making millions of what would otherwise have been individual and manual decisions. A digital world is electric, and a decarbonised world will be electric too. The underlying technological revolution that is digitalisation is complementary to the revolution in low-carbon generation.

Today's infrastructures are far from perfect and some are in an awful state. Aside from the carbon issues, we are facing many of the same challenges as the Victorians. Like them, we have new technologies, and we have big environmental and health problems to solve. Victorians had the sewage, the Great Stink of London, lots of water-borne diseases like cholera and typhoid, and they needed clean water for the growing urban populations. They needed to connect their manufacturing to world markets and for this they needed the canals and then the railways. We need a fibre network for communications; we need an electric road system; a railway fit for taking on what regional aviation delivers today; and above all we need an electricity system capable of harnessing active demand and generating decentralised low-carbon electricity.

Public finances would be transformed. There would be a carbon tax, and pollution charges would make up quite a lot of government revenues. They might be enough to reduce taxes on labour, like income tax and national insurance, and VAT too. Or they may be used to pay down our debts to the future generations, now augmented by the costs of Covid-19, and including the carbon in the atmosphere we will be bequeathing them.

There are many aspects to our individual lives which would be better in 2050 than they are now. A greener world is a healthier one. In addition to the health benefits from cleaner air and water, just being surrounded by more carbon-sequestering green improves mental health. It increases the scope for exercise and helps to address obesity. There would probably be less meat in many people's diets and that is a good thing, even if we don't all want or need to be vegetarian or vegan. Indoor pollution would be lower. Turn down the thermostat too, sleep better and generally be healthier.

Because there is a very real risk that climate change will not be cracked at the global level, temperatures may rise. They will anyway, since the carbon in the atmosphere is not going away for a very long time. This means we have to adapt. Despite the horror stories of the fate that will befall us as temperatures rise, even here there are quite a lot of things we can do which are no regrets. Creating new coastal wetlands increases resilience to flooding and improves biodiversity, as well as being very good at absorbing carbon. Seawalls, where they are needed, are quite cheap. A warmer UK needs less heating in winter, although probably more air conditioning in summer. The growing season will be longer, and people will spend more time outdoors.

What is not to like about these no regrets measures? Yet you might be thinking that this sounds too good to be true. What about the downsides of this transformation, and of course the costs? This is where realism comes in, and where it is important for our leaders to tell the truth. All the above is no regrets, and most of it will have to happen anyway. But just because it will have to happen does not make it cheap, and it will all need to be funded. The Victorian investments were a picnic in comparison with all of the above, and they took a century, while we have just 30 years to get to net zero. Remember too that this is net zero *consumption* and not just *production*. In this no regrets world, polluters have to pay to address the inefficiency of their pollution. Just because it is *no regrets* does not mean it will be *no cost* and therefore *no impact* on our standard of living.

No regrets requires us to get onto a sustainable economic growth path, and off the unsustainable GDP growth path. The point here is simple and fundamental: the reason a sustainable growth path is no regrets is because an unsustainable path will not be sustained. We will be forced onto a sustainable consumption path if we do nothing because we will reap the consequences of inaction.

The brutal fact that needs to be faced is that we are living beyond our environmental means, and that is very inefficient. Our consumption is too high: it is unsustainable even before we add on all that extra public spending, based on ever more debt, now promised by all main political parties in most developed countries and especially in the US and the UK. We need to save, to provide the funds to finance all the investment described above. Instead of saving virtually nothing, we may have to set aside more than 15 per cent of our income. That is a big hit, and it comes on top of the higher prices we will be paying through the carbon price for our carbon-intensive consumption. Some of this saving we will have to do anyway to pay for our pensions, which we chronically underfund, and for growing healthcare demands, especially in old age.

Even more brutal is the moral obligation to address the pollution we have already caused and which we are going to dump on the young and future generations, in addition to all the public and private debt we have piled up. This is where the easy fantasy of a green new deal drops away, and especially the assumption that if we have a carbon tax, the money can just be recycled back to us to spend. While it is true that it is better to tax 'bads' like pollution rather than 'goods' like labour, the first call on the carbon revenues should be to do the reparations. We have done the sort of damage that belligerent armies and air forces do in wars to their enemies. We ought to pay for it.

One way to do this is to put some of the carbon tax revenues into an environmental fund, and use this fund to make the investment in the infrastructures, pay down some of the costs of electrification of almost everything, pay for R&D, and compensate with offsets. In *Green and Prosperous Land*, I called this a Nature Fund. Expanded to incorporate climate change, the scope of the Fund could be correspondingly broader. With a carbon tax, the economics will be transformed anyway, as the relative price of carbon-intensive stuff will go up a lot, including carbon-intensive energy.

We have a stark choice. We could carry on being the selfish generation. In terms of the scale of the damage over the 30 wasted years, we are the most selfish generation in history. We could continue cheating the next and subsequent generations, fail to clean up our mess and carry on writing an ever-bigger mortgage on the future. Or we could face up to the sheer inefficiency of our economy, and get cracking on transforming the infrastructures, and introduce a comprehensive carbon price.

We can choose the second path sure in the knowledge that we at least are no longer causing further climate change, but only if we do it on a consumption basis. Net zero carbon production will not, contrary to what the CCC claims, do the job, and in some areas it might even make climate change worse. It is not true that 'By reducing emissions produced in the UK to zero, we also end our contribution to rising global temperatures.' If we choose to waste another 30 years, there will be damage. We will regret it, limping on with poor infrastructures, poor health and well-being, and low productivity, but not as much as the next generation will.

ACKNOWLEDGEMENTS

So much is now discussed, written and published about climate change that it is impossible to keep track of all the ideas and conversations that have influenced my understanding of the subject. If I leave out quite a few people, I can only apologise.

Of all the influences on my thinking, my colleague and friend Cameron Hepburn has been the most important. We do not always agree, but his knowledge and understanding of the issues, and his undimmed optimism that climate change can be solved, remain an inspiration to me. Together we have written on carbon border adjustments and climate policy, and we also founded Aurora Energy Research.

At Aurora, John Feddersen and Richard Howard have been particularly helpful. Thanks also to Asgeir Heimisson, Nayoung Kim and Hiren Mulchandani.

My interest in natural capital has carried over into climate change and, in particular, natural carbon sequestration. During my time as chair of the Natural Capital Committee, I have greatly benefited from trying out a number of the ideas in this book. Kathy Willis has been an inspiration, and Melanie Austen, Ian Bateman, Chris Collins, Paul Leinster and Colin Mayer have all helped in ways they are probably largely unaware of.

At Natural Capital Research, which Kathy Willis and I set up, Abigail Barker, Beccy Wilebore and Florian Zellweger have been big influencers on my thinking through how carbon sequestration works on the ground, how to think about baselines, and how to develop land use accordingly. Julian Metherell has recently added his probing and forensic style to this emerging market.

Over the last 20 years I have chaired the Energy Futures Group, where many of the climate change issues have been repeatedly discussed. I have greatly benefited from these discussions and conversations with Carl Arntzen, Alice Barrs, Rohit Bhrara, Michael Borrell, Katherine Collett, Tom Crotty, Kevin Dibble, Elena Giannakopoulou, Tom Glover, Angela Hepworth, Rhian Kelly, Matthew Knight, Pauline Lawson, Sinead Lynch, Andrew Mackintosh, Andrew Mennear, Steven Mills, John Moriarty, Thomas Mostyn, Nick Park, Tom Restrick, Matthew Setchell, Iain Smedley, Paul Spence, Brian Tilley, Sara Vaughan, Matt Willey and Matthew Wright.

More recently, I have chaired the Environment Group and again tried out several of the ideas set out in this book. I am grateful to Simon Bimpson, Owen Brennan, James Cooper, Adrian Dolby, David Elliot, Julie Fourcade, John Gilliland, Philip Gready, James Hall, Jo Harrison, Jacob Hayler, John Kimmance, Ed Mitchell, Darren Moorcroft, Simon Oates, Adam Read, Jake Rigg, Kitty Rose, Guy Thompson, Steven Thompson, John Varley, Iain Vosper, Pauline Walsh and Andrew Williams-Fry.

I have also made it my business to engage with the many companies that span the energy sector, and am particularly grateful for conversations with Chris Bennett, Salim Bensmail, James Bullock, Tomas David, Adrian Drummond, David Elliott, Fergal McNamara, Simone Rossi, Colin Skellett and Guy Thompson.

Across government, agencies and think tanks, conversations with Josh Buckland and Guy Newey have been particularly helpful. In preparing the Helm Review, the team at the Department for

Business, Energy and Industrial Strategy, led by Jeremy Allen, were outstanding. More generally at the Department I have benefited from conversations with Declan Burke and Dan Monzani, and with Treasury officials whom I shall not embarrass by naming. They know who they are. At Ofgem, Jonathan Brearley has been very helpful, as have David Joffe, Mike Thompson and Chris Stark on the Climate Change Committee.

The writing of the book has been made possible by the extremely generous support of Bjorn Saven. He has been the best of research funders, providing the financial support without any strings. I hope Bjorn feels his faith in my work has been justified by this book.

Many people have taken the trouble to comment on the first edition and have helped me improve this edition. They are too numerous to list, but they know who they are and I am most grateful to them all.

The Warden and Fellows of New College, Oxford, continue to provide me with the perfect environment in which to think and write. Hopefully this book might encourage the College to become an exemplar at net zero consumption.

On the practicalities, Kerry Hughes has been her usual brilliant and effective copy-editor, dealing with my scribbles and many changes. Jenny Vaughan has been brilliant too in helping to facilitate the whole process. Myles Archibald has again been the best of publishers: critical, supportive and good company. Thanks also to Hazel Eriksson and Helen Ellis at HarperCollins for their support and assistance.

Finally, as always, families bear the brunt of the concentrated effort needed to span 80,000 words. The book is dedicated to Sue Helm, who read and commented on earlier drafts, and to the growing Helm family – Oliver, Laura, and now Amelie and Jake.

ENDNOTES

PREFACE

1 Helm, D., *The Carbon Crunch: How We're Getting Climate Change Wrong – And How to Fix it*, London: Yale University Press, 2012; Helm, D., *Burn Out: The Endgame for Fossil Fuels*. London: Yale University Press, 2017; and Helm, D., 'The Cost of Energy Review' (the Helm Review), report prepared for the Department for Business, Energy and Industrial Strategy, October 2017.

2 Climate Change Committee, 'Net Zero: The UK's Contribution to Stopping Global Warming', May 2019, p 8.

3 COP26 is the 26th United Nations Climate Change Conference of the Parties, to be hosted in Glasgow in November 2021.

4 Climate Change Committee, 'The Sixth Carbon Budget: The UK's Path to Net Zero', December 2020.

5 Helm, D., *Natural Capital: Valuing the Planet*. London: Yale University Press, 2015. Helm, D., *Green and Prosperous Land: A Blueprint for Rescuing the British Countryside*. London: HarperCollins, 2019.

6 Department for Environment, Food and Rural Affairs, 'A Green Future: Our 25 Year Plan to Improve the Environment', January 2018.

INTRODUCTION

1 See World Meteorological Organization, 'Greenhouse Gas Bulletin: The State of Greenhouse Gases in the Atmosphere Based on Global Observations through 2019', November 2020, https://library.wmo.int/index.php?lvl=notice_display&id=21795#.YEt3SWj7Q55.

2 Pfeiffer, A., Millar, R., Hepburn, C. and Beinhocker, E., 'The "2°C capital

stock" for electricity generation: Committed cumulative carbon emissions from the electricity generation sector and the transition to a green economy', *Applied Energy*, 179, 2016, pp 1395–1408.

3 Gallagher, K. S., Bhandary. R., Narassimhan, E. and Nguyen, Q. T., 'Banking on coal? Drivers of demand for Chinese overseas investments in coal in Bangladesh, India, Indonesia and Vietnam', *Energy, Research & Social Science*, 71, January 2021.

4 The largest expenditure item for the average UK household in 2018 was transport (14 per cent). Adding housing, fuel and power gets to nearly 50 per cent. See www.ons.gov.uk/peoplepopulationandcommunity/personaland-householdfinances/expenditure/bulletins/familyspendingintheuk/financialyearending2018.

1
NO PROGRESS

1 'What we are now doing to the world, by degrading the land surfaces, by polluting the waters and by adding greenhouse gases to the air at an unprecedented rate – all this is new in the experience of the earth. It is mankind and his activities which are changing the environment of our planet in damaging and dangerous ways . . . Mr President, the evidence is there. The damage is being done. What do we, the International Community, do about it?' Speech by Margaret Thatcher to United Nations General Assembly (Global Environment), 8 November 1989, New York, https://www.margaretthatcher.org/document/107817.

2 United Nations, 'United Nations Framework Convention on Climate Change', 1992, https://unfccc.int/resource/docs/convkp/conveng.pdf.

3 United Nations, 'Kyoto Protocol to the United Nations Framework Convention on Climate Change', 1998, https://unfccc.int/resource/docs/convkp/kpeng.pdf; United Nations, 'Paris Agreement', 2015, https://unfccc.int/sites/default/files/english_paris_agreement.pdf

4 For explanatory ease, I refer to carbon and CO_2 throughout the book as a shorthand for greenhouse gases, except where explicitly stated. The important differences between the gases are glossed over in order to avoid complicating the overall narrative.

5 Figure 1 sourced from Dr Pieter Tans, NOAA/GML (www.esrl.noaa.gov/gmd/ccgg/trends/) and Dr Ralph Keeling, Scripps Institution of Oceanography (scrippsco2.ucsd.edu/). Retrieved from www.esrl.noaa.gov/gmd/ccgg/trends/data.html.

6 See World Meteorological Organization, 'Greenhouse Gas Bulletin: The State of Greenhouse Gases in the Atmosphere Based on Global Observations through 2019', November 2020, https://library.wmo.int/index.php?lvl=notice_display&id=21795#.YEt3SWj7Q55.

7 Figure 2 sourced from Ritchie, H. and Roser, M., 'Atmospheric CO2 Concentration', 2019. Underlying data from EPICA Dome C CO2 record (2015) and NOAA (2018). Published online at OurWorldInData.org. Retrieved from https://ourworldindata.org/co2-and-other-greenhouse-gas-emissions.

8 Figure 3 sourced from NOAA Climate.gov. Retrieved from www.climate.gov/news-features/understanding-climate/climate-change-atmospheric-carbon-dioxide. Lüthi, D., Le Floch, M., Bereiter, B., Blunier, T., Barnola, J.-M., Siegenthaler, U., Raynaud, D., Jouzel, J., Fischer, H., Kawamura, K. and Stocker, T. F., 'High-resolution carbon dioxide concentration record 650,000–800,000 years before present', *Nature*, 453, 2008, pp. 379–382. doi:10.1038/nature06949.

9 Figure 4 sourced from Ritchie, H. and Roser, M., 'Global Primary Energy Consumption by Source', 2020. Underlying data from Smil, V. (2017) and BP Statistical Review of World Energy. Published online at OurWorldInData.org. Retrieved from https://ourworldindata.org/grapher/global-energy-substitution?country=~OWID_WRL.

10 See International Organization of Motor Vehicle Manufacturers, '2018 Production Statistics', www.oica.net/category/production-statistics/2018-statistics/.

11 See Statista, 'International seaborne trade carried by container ships from 1980 to 2017', www.statista.com/statistics/253987/international-seaborne-trade-carried-by-containers/.

12 See the International Energy Agency's (IEA's) projection of oil demand by sector through to 2030 at www.iea.org/petrochemicals/.

13 Widger, P. and Haddad, A. M., 'Evaluation of SF_6 Leakage from Gas Insulated Equipment on Electricity Networks in Great Britain', *Energies*, 11, 2037, August 2018.

14 Figure 5 based on IEA, 'Coal Information Overview: Statistics Report – July 2020', IEA, Paris, 2020. Data retrieved from https://www.iea.org/reports/coal-information-overview.

15 Sourced from Ritchie, H. and Roser, M., 'Coal Production', 2020. Underlying data from BP Statistical Review of World Energy and Shift Data Portal. Published online at OurWorldInData.org. Retrieved from https://ourworldindata.org/fossil-fuels.

16 International Energy Agency, 'World Energy Outlook 2018', 2018.

17 Peak coal has been frequently predicted, but never realised. William Stanley Jevons famously got it all very wrong in the nineteenth century when he warned that, as England ran out, it would return to its former littleness. Jevons, W. S., *The Coal Question: An Inquiry Concerning the Progress of the Nation, and the Probable Exhaustion of Our Coal-mines*, London: Dodo Press, 2008 edition.

18 These are described in Helm, D., *Burn Out: The Endgame for Fossil Fuels*. London: Yale University Press, 2017.

19 See Statista, 'Major foreign holders of U.S. treasury securities as of June 2019', www.statista.com/statistics/246420/major-foreign-holders-of-us-treasury-debt/.

20 See www.iea.org/topics/coal/statistics/.

21 See Chatzky, A. and McBride, J., 'China's Massive Belt and Road Initiative', Council on Foreign Relations, last updated 21 May 2019, www.cfr.org/backgrounder/chinas-massive-belt-and-road-initiative?gclid=EAIaIQobChMIyrP80cqA5gIVBrDtCh0UGA_VEAAYAyAAEgIV1PD_BwE.

22 Figure 7 sourced from World Bank national accounts data, and OECD National Accounts data files. The World Bank, https://data.worldbank.org/indicator/NY.GDP.MKTP.CD?locations=CN.

23 Table 1 sourced from Ritchie, H. and Roser, M., 'CO_2 Emissions', 2020. Published online at OurWorldInData.org. Retrieved from https://ourworldindata.org/co2-emissions and https://ourworldindata.org/per-capita-co2.

24 According to BP Energy Outlook 2019, coal's share in India's primary energy consumption will decline from 56 per cent in 2017 to 48 per cent in 2040. BP, 'BP Energy Outlook: 2019 Edition', BP plc, 2019.

25 Figure 8 sourced from Union of Concerned Scientists, based on Earth Systems Science Data 11, 1738-1838, 2019, https://www.ucsusa.org/resources/each-countrys-share-co2-emissions.

26 Figure 9 sourced from Ritchie, H. and Roser, M., 'Renewable Energy Consumption, World', 2020. Underlying data from Vaclav Smil (2017) and BP, 'Statistical Review of Global Energy', 2020. Published online at OurWorldInData.org. Retrieved from https://ourworldindata.org/grapher/renewable-energy-consumption?country=~OWID_WRL.

27 International Energy Outlook, 'Offshore Wind Outlook 2019', World Energy Outlook Special Report, October 2019.

28 In its enthusiasm to reinvent itself as a renewables champion, the IEA forecasts that offshore wind will be cost-competitive with fossil fuels by 2030 – conveniently forgetting how far fossil fuel production costs and market

prices could also fall over the same period, and leaving out the additional network investments required. International Energy Outlook, 'Offshore Wind Outlook 2019', World Energy Outlook Special Report, October 2019.

29 See Staedter, T., 'Big Mammals Evolved Thanks to More Oxygen', *Scientific American*, 3 October 2005, www.scientificamerican.com/article/big-mammals-evolved-thank/.

30 See De Leon, T., 'Burning Indonesian peat causes haze in Singapore', MIT News, 21 November 2018, http://news.mit.edu/2018/mit-researchers-peat-burning-sumatra-causes-severe-haze-singapore-1121.

31 See United States Environmental Protection Agency, 'Global Greenhouse Gas Emissions Data', www.epa.gov/ghgemissions/global-greenhouse-gas-emissions-data. The UK figure is 10 per cent, although it leaves out quite a lot of soil- and peat-related losses. The actual share is likely be considerably higher once all the factors are taken into account.

2
THE ROAD TO GLASGOW

1 For an analysis of the history of the UN, see Kennedy, P. *The Parliament of Man: The Past, Present and Future of the United Nations*. London: Penguin, 2006.

2 See Barrett, S., *Environment and Statecraft: The Strategy of Environmental Treaty-making*. Oxford: Oxford University Press 2005; and Victor, D. G., *Global Warming Gridlock: Creating More Effective Strategies for Protecting the Planet*. Cambridge: Cambridge University Press, 2011.

3 Fukuyama, F., *The End of History and the Last Man*. Hamondsworth, Penguin, 1989.

4 Stern, N., *The Economics of Climate Change: The Stern Review*, HM Treasury. Cambridge: Cambridge University Press, January 2007.

5 For an analysis, see Baltensperger, M. and Dadush, U., 'The European Union–Mercosur Free Trade Agreement: Prospects and risks', Policy Contribution, 11 September 2019, https://bruegel.org/wp-content/uploads/2019/09/PC-11_2019.pdf.

6 United Nations, 'Our Common Future: Report of the World Commission on Environment and Development', The Brundtland Report, 1987.

7 Hume, D., *Dialogues Concerning Natural Religion*. 1779.

8 For the impacts of falling oil prices on the Middle East and Russia, respectively, see chapters 5 and 6 of Helm, D., *Burn Out: The Endgame for Fossil Fuels*. London: Yale University Press, 2017.

9 See Victor, D. G., *The Collapse of the Kyoto Protocol and the Struggle to Slow Global Warming*. Princeton: Princeton University Press, 2004.

10 For the UN's optimistic explanation of what was agreed at Paris see https://unfccc.int/process-and-meetings/the-paris-agreement/what-is-the-paris-agreement.

3
GOING IT ALONE

1 European Commission, 'The European Green Deal', COM(2019) 640 final, 11 December 2019. See also European Commission, 'Proposal for a Regulation of the European Parliament and of the Council establishing the framework for achieving climate neutrality and amending Regulation (EU) 2018/1999 (European Climate Law)', COM(2020) 80 final, 2020/0036 (COD), 4 March 2020; European Commission, 'European Climate Pact', https://ec.europa.eu/clima/policies/eu-climate-action/pact_en; and European Commission, 'A new Industrial Strategy for a globally competitive, green and digital Europe', 10 March 2020.

2 David Hume famously addressed what became known as the 'is-ought' problem in philosophy. See Hume, D., *An Enquiry Concerning the Principles of Morals*. 1777. For a discussion, see MacIntyre, A. C., 'Hume on "Is" and "Ought"', *The Philosophical Review*, 68(4), October 1959, pp 451–468.

3 Some utilitarians get around this problem by arguing for a rules-based approach: we should adopt rules that have the effect of maximising aggregate utility. The problem with this is that any rule could be reinterpreted as if it is consistent with utilitarianism, but this does not make it so. On rules-based utilitarianism, see Smart, J. J. C. and Williams, B., *Utilitarianism: For and Against*. New York: Cambridge University Press, 1973.

4 European Commission, 'Proposal for a Regulation of the European Parliament and of the Council establishing the framework for achieving climate neutrality and amending Regulation (EU) 2018/1999 (European Climate Law)', COM(2020) 80 final, 2020/0036 (COD), 4 March 2020.

5 Directive 2009/28/EC of the European Parliament and of the Council of 23 April 2009 on the promotion of the use of energy from renewable sources and amending and subsequently repealing Directives 2001/77/EC and 2003/30/EC.

6 See 'Internal energy market', Fact Sheet on the European Union, www.europarl.europa.eu/factsheets/en/sheet/45/internal-energy-market.

7 See Helm, D., 'European Energy Policy', in Jones, E., Menon, A. and Weatherill, S. (eds), *The Oxford Handbook of the European Union*. Oxford: Oxford University Press, 2012, chapter 39.

8 On the EU as a whole, see Becqué, R., Dubsky, E., Hamza-Goodacre, D. and Lewis, M., 'Europe's Carbon Loophole', September 2017. On Denmark's performance see Energistyrelsen, 'Small increase in energy consumption in 2018', press release, April 2019, https://presse.ens.dk/pressreleases/lille-stigning-i-energiforbruget-i-2018-2856269. For a comment, see CleanTechnica, 'New Report On Energy Consumption In Denmark – Carbon Emission Increased In 2018', 8 April 2019, https://cleantechnica.com/2019/04/08/new-report-on-energy-consumption-in-denmark-carbon-emission-increasein-2018/.

9 European Commission, 'Biofuels', https://ec.europa.eu/energy/en/topics/renewable-energy/biofuels/overview.

10 For a review, see Muûls, M., Colmer, J., Martin, R. and Wagner, U. J., 'Evaluating the EU Emissions Trading System: Take it or leave it? An assessment of the data after ten years', Grantham Institute Briefing paper No. 21, October 2016.

11 See European Court of Auditors, 'The integrity and implementation of the EU ETS', Special Report, 2015.

12 This is set out in detail in the Helm Review.

13 See Helm, D. 'Agriculture after Brexit', *Oxford Review of Economic Policy*, 33(1), 2017, pp 124–133; and chapter 4 of Helm, D., *Green and Prosperous Land: A Blueprint for Rescuing the British Countryside*. London: HarperCollins, 2019.

14 Federal Ministry for Economic Affairs and Energy, 'Commission on Growth, Structural Change and Employment: Final Report', January 2019. It might in due course try to go a bit faster.

4
LIVING WITHIN OUR ENVIRONMENTAL MEANS

1 The fashion for 'rewilding' confuses the advocacy of a particular management technique in particular circumstances with a back-to-nature romanticism.

2 Technically, this includes rules in respect to the depletion charges of renewable and non-renewable natural resources. For the classic treatment, see Dasgupta, P. S. and Heal, G. M., *Economic Theory and Exhaustible Resources*. Cambridge: James Nisbet and Cambridge University Press, 1979.

3 This approach became known as 'shock therapy' and gained further traction after the global financial crash in 2007/08. For the classic article on transition,

see Lipton, D. and Sachs, J. D., 'Creating a Market Economy in Eastern Europe: The Case of Poland', Brookings Papers on Economic Activity, 1, 1990.

4 See Helm, D., 'The Economic Impacts of the Coronavirus', *Environmental and Resource Economics*, (76), 2020, pp 21–38.

5 Helm, D., 'Net Carbon Gain', February 2021, http://www.dieterhelm.co.uk/natural-capital/environment/net-carbon-gain/.

6 Dasgupta, P., *Time and the Generations*. New York: Columbia University Press, 2019. See also Helm, D., 'The Sustainable Borders of the State', *Oxford Review of Economic Policy*, (27)4, 2011, pp 517–535.

7 Among the more recent contributions, see Jackson, T., *Prosperity without Growth*, Second Edition. New York: Routledge, 2017. See also Hamilton, C., *Growth Fetish*, London: Allen & Unwin, 2003.

8 See Malthus, T., *Essay on the Principle of Population*. London: Pelican Books, 1970; Ehrlich, P. R., *The Population Bomb: Population Control or Race to Oblivion?*. New York: Ballantine Books, 1968; and Meadows, D. H., Meadows, D. L., Randers, J. and Behrens, W., 'The Limits to Growth: A Report for the Club of Rome's Project on the Predicament of Mankind'. New York: Universe Books, 1972.

9 See Coyle, D., *GDP: A Brief but Affectionate History*. Princeton: Princeton University Press.

10 See Helm, D. 'Rethinking the economic borders of the state – ownership, assets, and competition', *Oxford Review of Economic Policy*, 31(2), 2015, pp 168–185.

11 See, for example, https://rebellion.earth/act-now/resources/citizens-assembly/.

12 John Stuart Mill, for example, recommended that the vote should be extended only to those with sufficient education. See Miller, J. J., 'J. S. Mill on Plural Voting, Competence and Participation', *History of Political Thought*, 24(4), 2003, pp 647–667. See also Mulligan, T., 'Plural Voting for the Twenty-First Century', *The Philosophical Quarterly*, 68(271), September 2017, pp 286–306.

13 Jonathan Porritt, for example, wrote a deeply sceptical book, *Playing Safe: Science and the Environment*. Porritt, J., *Playing Safe: Science and the Environment*. London: Thames and Hudson, 2000.

14 For a history, see Morgan, K. O., *Labour in Power, 1945–1951*. Oxford: Oxford University Press, 1984, pp 49–50.

15 See Helm, D., *Energy, the State and the Market: British Energy Policy Since 1979*. Oxford: Oxford University Press, 2004.

16 See the discussion in the introduction.

17 The irony of this perfect competition model is that if all information is

perfect, then the market is not actually needed. The State could make perfect decisions too.

18 See Berlin, I., *The Crooked Timber of Humanity: Chapters in the History of Ideas*. London: Pimlico, 2013.

5
THE PRICE OF CARBON

1 The discussion here focuses on carbon in the narrowest sense. In due course, pricing could be extended to the other greenhouse gases, notably methane.

2 For a broad discussion of the ethical issues, see Broome, J., *Climate Matters: Ethics in a Warming World*. New York: W. W. Norton & Company Inc., 2012.

3 HM Treasury, 'Net Zero Review: Interim Report', December 2020, https://www.gov.uk/government/publications/net-zero-review-interim-report.

4 They are not strictly equal since the income effect will vary: whether the tax leaves consumers poorer or not determines how much they will have to spend – including on carbon intensive products.

5 See Hepburn, C., 'Regulation by prices, quantities, or both: a review of instrument choice', *Oxford Review of Economic Policy*, 22(2), 2006, pp 226–247.

6 See the Energy White Paper and the preceding Treasury consultation paper which provided a much more balanced assessment. Department for Business, Energy and Industrial Strategy, 'The Energy White Paper: Powering our Net Zero Future', December 2020, https://www.gov.uk/government/publications/energy-white-paper-powering-our-net-zero-future; HM Treasury, 'Carbon Emissions Tax: Consultation', July 2020, https://assets.publishing.service.gov.uk/government/uploads/system/uploads/attachment_data/file/902737/Carbon_Emissions_Tax_-_consultation.pdf.

7 See, for example, Nordhaus, W. D., 'Revisiting the social cost of carbon', PNAS, 114(7), February 2017, pp 1518–1523.

8 See Nordhaus, W., *The Climate Casino: Risk, Uncertainty, and Economics for a Warming World*. Yale University Press, 2013, pp 28–30. For a critical review, see Ackerman, F., DeCanio, S. J., Howarth, R. B. and Sheeran, K., 'Limitations of integrated assessment models of climate change', *Climatic Change*, 95, 2009, pp 297–315.

9 These objections are examined in more detail in Helm, D., Hepburn, C. and Ruta, G., 'Trade, Climate Change, and the Political Game Theory of Border Carbon Adjustments', *Oxford Review of Economic Policy*, 28(2), July 2012. See

also Lowe, S., 'Should the EU tax imported CO_2?', Centre for European Reform, September 2019.

10 See again Helm, D., Hepburn, C. and Ruta, G., 'Trade, Climate Change, and the Political Game Theory of Border Carbon Adjustments', *Oxford Review of Economic Policy*, 28(2), July 2012.

11 The protection of the farmers' red diesel subsidy in the March 2020 budget, after sustained lobbying, is another example.

12 The Fuel Duty freeze in the March 2021 Budget reinforces the dominance of short-term considerations over the medium-term decarbonisation objective.

13 See Fullerton, D., Leicester, A. and Smith, S. 'Environmental taxes', in Institute for Fiscal Studies, *Reforming the Tax System for the 21st Century: The Mirrlees Review*. Oxford: Oxford University Press, 2011.

14 As an example, see, Beiser-McGrath, L. F. and Bernauer, T., 'Could revenue recycling make effective carbon taxation politically feasible?', *Science Advances*, 5(9), 18 September 2019.

15 For how this might work, see 'A Nature Fund', chapter 10 in Helm, D., *Green and Prosperous Land: A Blueprint for Rescuing the British Countryside*. London: HarperCollins, 2019, and Helm, D., 'The ownership and funding of natural capital: The case for trusts and a public natural capital fund', in *International Journal of Public Policy*, 15:1/2, 2019.

6
NET ZERO INFRASTRUCTURES

1 Helm, D., 'The new broadband utility and the Openreach debate', January 2016, www.dieterhelm.co.uk/regulation/communications/the-new-broadband-utility-and-the-openreach-debate/.

2 The Norwegian electricity network is run by state-owned Statnett, which has already completed its roll-out of smart meters.

3 For example, when he was prime minister, David Cameron committed to the USO for broadband back in 2012. See Gov.uk, 'PM speech on infrastructure: Prime Minister's speech on national infrastructure was delivered at the Institute of Civil Engineering', 19 March 2012, www.gov.uk/government/speeches/pm-speech-on-infrastructure.

4 HM Treasury, 'National Infrastructure Strategy', November 2020, https://www.gov.uk/government/publications/national-infrastructure-strategy.

5 HM Government, 'The Ten Point Plan for a Green Industrial Revolution',

November 2020, https://www.gov.uk/government/publications/the-ten-point-plan-for-a-green-industrial-revolution; Department for Business, Energy and Industrial Strategy, 'The Energy White Paper: Powering our Net Zero Future', December 2020, https://www.gov.uk/government/publications/energy-white-paper-powering-our-net-zero-future.

6 The ambitious objective of switching from road traffic to railways featured strongly in the Department of the Environment, Transport and the Regions (DETR) 10-year transport plan, 'Transport 2010'. The secretary of state at the time, John Prescott, stated in the House of Commons that 'Large-scale investment in the upgrading and expansion of the network will allow 50% more passengers to travel by train more quickly and comfortably, in greater safety, more punctually, between more attractive stations. Investment in infrastructure will encourage, by our estimate, an increase of 80% in goods carried by rail.'

7 Department for Transport, 'Decarbonising Transport: Setting the Challenge', March 2020.

8 Helm, D. 'HS2: a conclusion in search of a rationale', September 2019, www.dieterhelm.co.uk/regulation/transport/hs2-a-conclusion-in-search-of-a-rationale/.

9 See the Airports Commission, 'Airports Commission: Final Report', July 2015.

10 This is known as the Jevons Paradox after the nineteenth-century English economist William Stanley Jevons, referred to in chapter 1: more efficiency leads to higher, not lower, demand, as it has for the last 200 years since those early very inefficient steam engines were used to pump water from coal mines.

11 One proposal is Desertec, which failed, but nevertheless identified the ambitious opportunity which might be subsequently pursued. See www.euractiv.com/section/trade-society/news/desertec-abandons-sahara-solar-power-export-dream/.

12 On the CEGB, see Helm, D., *Energy, the State and the Market: British Energy Policy Since 1979*. Oxford: Oxford University Press, 2004.

13 On system operators and their regulation, see Helm, D., 'The Systems Regulation Model', February 2019, www.dieterhelm.co.uk/regulation/regulation/the-systems-regulation-model/.

14 See also Ofgem, 'Review of GB energy system operation', January 2021, https://www.ofgem.gov.uk/system/files/docs/2021/01/ofgem_-_review_of_gb_energy_system_operation_0.pdf.

7
NATURAL SEQUESTRATION, OFFSETTING, AND CARBON CAPTURE AND STORAGE

1 On the role of trees, see Bastin, J-F., Finegold, Y., Mollicone, D., Rezende, M., Routh, D., Zohner, C. M. and Crowther, T. W., 'The global tree restoration potential', *Science*, 365, 2019, pp 76–79. The study suggests that 1.2 trillion trees cost US$300 billon, and 11 per cent of global land area. On peat bogs, see Garnett, S., Selvidge, J., Westerberg, S. and Thompson, P., 'RSPB Geltsdale – a case study of upland management', *British Wildlife*, 30(6), August 2019, pp 409–417.

2 Indeed, this nascent technology is already being developed, albeit on a very small scale. See, for example, www.climeworks.com.

3 In 2019 Drax launched a bioenergy CCS pilot to trial the capture of CO_2 produced from the combustion of biomass.

4 See Hepburn, C., Adlen, E., Beddington, J., Carter, E. A., Fuss, S., MacDowell, N., Minx, J. C., Smith, P. and Williams, C. K., 'The technological and economic prospects for CO_2 utilization and removal', *Nature*, 575, 2019, pp 87–97.

5 See Goldthorpe, S. 'Potential for Very Deep Ocean Storage of CO_2 Without Ocean Acidification: A Discussion Paper', *Energy Procedia*, 114, July 2017, pp 5417–5429. See also British Geological Survey, 'How can CO_2 be stored?', www.bgs.ac.uk/discoveringGeology/climateChange/CCS/howCanCo2Be Stored.html.

6 See Woodland Trust, 'The Northern Forest', www.woodlandtrust.org.uk/about-us/what-we-do/we-plant-trees/the-northern-forest/.

7 See Natural Capital Committee, 'Advice on using nature based interventions to reach net zero greenhouse gas emissions by 2050', April 2020; and Climate Change Committee, 'Land use: Policies for a Net Zero UK', January 2020.

8 Department for Environment, Food and Rural Affairs, 'A Green Future: Our 25 Year Plan to Improve the Environment', January 2018.

9 The CCC Net Zero report recommends that 'the target should be achieved through UK domestic effort, without relying on international carbon units (or "credits")' (p. 15).

10 This is a key challenge for the Clean Development Mechanism, the UN transfer mechanism for money from developed to developing countries, put in place under the Kyoto Protocol. See United Nations, 'The Clean Development Mechanism', https://unfccc.int/process-and-meetings/the-kyoto-protocol/mechanisms-under-the-kyoto-protocol/the-clean-development-mechanism.

11 For a useful guide, see Berners-Lee, M., *How Bad Are Bananas? The Carbon Footprint of Everything*, Updated and Expanded, London: Profile Books, 2020.

12 Interview on BBC1 *Panorama*, 'Can Flying Go Green?', broadcast 11 November 2019. Ryanair also has a project in Uganda on fuel stoves.

8

AGRICULTURE: GREEN, PROSPEROUS AND LOW-CARBON

1 For a detailed report about how human activity – and agriculture in particular – is affecting nature and biodiversity in the UK, see Hayhow, D. B., Eaton, M. A., Stanbury, A. J., Burns, F., Kirby, W. B., Bailey, N., Beckmann, B., Bedford, J., Boersch-Supan, P. H., Coomber, F., Dennis, E. B., Dolman, S. J., Dunn, E., Hall, J., Harrower, C., Hatfield, J. H., Hawley, J., Haysom, K., Hughes, J., Johns, D. G., Mathews, F., McQuatters-Gollop, A., Noble, D. G., Outhwaite C. L., Pearce-Higgins, J. W., Pescott, O. L., Powney, G. D. and Symes, N., 'The State of Nature 2019', The State of Nature Partnership, October 2019.

2 'Consumer-facing policies should be used to support shifts to healthier diets with lower beef, lamb and dairy consumption. This would allow changes in UK land use without increasing reliance on imports. Forest cover should increase from 14% of UK land to 17% by 2050.' Climate Change Committee, 'Net Zero: The UK's contribution to stopping global warming', May 2019, p 35. This was updated and further elaborated on in the CCC's Sixth Carbon Budget. Climate Change Committee, 'The Sixth Carbon Budget: The UK's Path to Net Zero', December 2020.

3 NFU, 'Achieving net zero – meeting the climate change challenge', www.nfuonline.com/news/latest-news/achieving-net-zero-meeting-the-climate-change-challenge/.

4 See Natural Capital Committee, 'Advice on using nature based interventions to reach net zero greenhouse gas emissions by 2050', April 2020.

5 This was all pointed out in the 2017 Helm Review.

6 See Helm, D. 'Agriculture after Brexit', *Oxford Review of Economic Policy*, 33(1), 2017, pp 124–133; and Centre for Rural Economy, 'Brexit: how might UK agriculture thrive or survive?', Note No 7, School of Natural and Environmental Sciences, Newcastle University, August 2018.

7 For background on salmon farming, see Rural Economy and Connectivity Committee, 'Salmon farming in Scotland', Scottish Parliament Paper 432, November 2018.

8 See chapter 3, 'Defining the aggregate natural capital rule', in Helm, D., *Natural Capital: Valuing the Planet*. London: Yale University Press, 2015.
9 Mendel, G., *Experiments on Plant Hybridisation*, 1865. See also Tudge, C., *In Mendel's Footnotes: An Introduction to the Science and Technologies of Genes and Genetics from the Nineteenth Century to the Twenty-Second*. Vintage, 2002.
10 See, for example, Harper Adams University's Hands Free Hectare project at www.harper-adams.ac.uk/research/project/197/hands-free-hectare.
11 See the Royal Society, 'Genetic technologies'. Available at https://royalsociety.org/topics-policy/projects/genetic-technologies/.
12 For a radical view of the options, see Tubb, C. and Seba, T., 'Rethinking Food and Agriculture 2020–2030: The Second Domestication of Plants and Animals, the Disruption of the Cow, and the Collapse of Industrial Livestock Farming', RethinkX, September 2019.
13 For a series of insights into insect farming, see www.sciencedirect.com/topics/agricultural-and-biological-sciences/insect-farming.
14 The Hutton Institute has an indoor vertical farming collaborative project with Intelligent Growth Solutions. www.hutton.ac.uk/news/james-hutton-institute-congratulates-intelligent-growth-solutions-achieving-%C2%A354m-funding-boost.
15 Department for Environment, Food and Rural Affairs, 'A Green Future: Our 25 Year Plan to Improve the Environment', January 2018.

9
REINVENTING TRANSPORT

1 Johnson, S. and Boswell, J., *A Journey to the Western Islands of Scotland and the Journal of a Tour to the Hebrides*, edited by P. Levi. Penguin Classics, 1984.
2 The classic history of the oil industry is Yergin, D., *The Prize: The Epic Quest for Oil, Money and Power*. New York: Free Press, 1991.
3 'Producing an electric vehicle contributes, on average, twice as much to global warming potential and uses double the amount of energy than producing a combustion engine car.' World Economic Forum, 'Batteries can be part of the fight against climate change – if we do these five things', www.weforum.org/agenda/2017/11/battery-batteries-electric-cars-carbon-sustainable-power-energy/.
4 On the dimensions of transport demand, see papers in OECD 2003 European Conference of Transport Ministers, 'Managing the Fundamental Drivers of Transport Demand', www.itf-oecd.org/sites/default/files/docs/03demand.pdf.

See also, Department for Transport, 'Decarbonising Transport: Setting the Challenge', March 2020.

5 See also, Department for Transport, 'Decarbonising Transport: Setting the Challenge', March 2020.

6 See Alarco, J. and Talbot, P., 'The history and development of batteries', Phys. org, https://phys.org/news/2015-04-history-batteries.html.

7 The battery-swapping option was developed by the Better Place company in Israel. It went bankrupt in 2013. For applications to buses and public transport, and notably in China, see Li, W., Li, Y., Deng, H. and Bao, L., 'Planning of Electric Public Transport System under Battery Swap Mode', *Sustainability*, 10, 2528, July 2018.

8 In response to the National Grid estimates of a significant additional capacity requirement, see Morstyn, T., Teytelboym, A. and McCulloch, M. D., 'Designing Decentralized Markets for Distribution System Flexibility', *IEEE Transactions on Power Systems*, 34(3), May 2019. National Grid's modelling is presented in, 'Future Energy Scenarios', 2017, http://fes.nationalgrid.com/. See also Catapult Energy Systems, 'Preparing UK Electricity Networks for Electric Vehicles', Report, 2018.

9 On ammonia, see the Royal Society, 'Ammonia in a Net-zero Carbon Future: A carbon free fuel and energy store', 2020.

10 On future scenarios for oil prices, see Helm, D., *Burn Out: The Endgame for Fossil Fuels*. London: Yale University Press, 2017.

11 A takeover of British Steel by the Chinese company Jingye would not invalidate this point, which is about location and not ownership. However, if Jingye imports raw steel to the UK plants and then finishes these products, the carbon border tax should apply to what is imported.

12 On average, greenhouse gas emissions from corn ethanol are 34 per cent lower than from petrol when including land-use change emissions, and 44 per cent lower when excluding land-use change emissions. See Wang, M., Han, J., Dunn, J. B., Cai, H. and Elgowainy, A., 'Well-to-wheels energy use and greenhouse gas emissions of ethanol from corn, sugarcane and cellulosic biomass for US use', *Environmental Research Letters*, 7, 2012, pp 1–13.

13 The new energy crops extend to maize and rye grass in the UK for anaerobic digesters. Roughly 30 per cent of maize gets used in this way, notwithstanding the multiple environmental costs.

14 Helm, D., 'Regulatory Reform, Capture, and the Regulatory Burden', *Oxford Review of Economic Policy*, 22(2), Summer 2006, pp 169–185.

15 These are dealt with by the UN's International Maritime Organisation, which now has a 50 per cent emissions reduction target by 2050, and the UN's International Civil Aviation Organisation with its Carbon Offsetting and Reduction Scheme for International Aviation agreement.

16 On the pollution from cruise holidays, see www.transportenvironment.org/press/luxury-cruise-giant-emits-10-times-more-air-pollution-sox-all-europe%E2%80%99s-cars-%E2%80%93-study.

10
THE ELECTRIC FUTURE

1 HM Government, 'The Ten Point Plan for a Green Industrial Revolution', November 2020, https://www.gov.uk/government/publications/the-ten-point-plan-for-a-green-industrial-revolution; Department for Business, Energy and Industrial Strategy, 'The Energy White Paper: Powering our Net Zero Future', December 2020, https://www.gov.uk/government/publications/energy-white-paper-powering-our-net-zero-future.

2 Department for Business, Energy and Industrial Strategy, 'Implementing the End of Unabated Coal by 2025: Government response to unabated coal closure consultation', January 2018.

3 The 2014 forecast by the Department of Energy and Climate Change (DECC – the predecessor of the Department for Business, Energy and Industrial Strategy), ahead of the crash in oil prices, suggested a continuing rising wholesale electricity price, and underpinned the £92/megawatt hour in the Hinkley Contract for Difference (CfD), indexed over the 35-year contract life. In expecting the wholesale price to keep on rising above the CfD value, DECC expected Hinkley to keep customers 'in-the-money'.

4 On the system regulation model, see Helm, D., 'The Systems Regulation Model', February 2019, www.dieterhelm.co.uk/regulation/regulation/the-systems-regulation-model/. On the extension of the RAB model to nuclear, see Helm, D., 'The Nuclear RAB Model', June 2018, www.dieterhelm.co.uk/energy/energy/the-nuclear-rab-model/.

5 The gradual demise of the wholesale market in electricity is discussed in further detail in the Helm Review, and chapters 10 and 11 of Helm, D., *Burn Out: The Endgame for Fossil Fuels*. London: Yale University Press, 2017.

6 National Grid's detailed report on how the power cut happened, and the system responses to the regulator, is available at www.ofgem.gov.uk/publications-and-updates/ofgem-has-published-national-grid-electricity-

system-operator-s-technical-report. For an analysis of the underlying causes, see Helm, D., 'Power cuts and how to avoid them', August 2019, www.dieterhelm.co.uk/energy/energy/power-cuts-and-how-to-avoid-them/.

7 A good example is the Community Centre in North Uist in the Outer Hebrides. It has a single turbine, and next to it an electric car charging point. The contrast with the ongoing cutting of peat nearby for household heating and even cooking stoves is stark.

8 For a history of the postwar period and the success of the CEGB in meeting the rapid growth of electricity demand, see Helm, D., *Energy, the State and the Market: British Energy Policy Since 1979*, Oxford: Oxford University Press, 2004.

9 On EMR, see Department of Energy and Climate Change, 'Implementing Electricity Market Reform (EMR): Finalised policy positions for implementation of EMR', policy paper, June 2014. See also www.ofgem.gov.uk/electricity/wholesale-market/market-efficiency-review-and-reform/electricity-market-reform-emr.

10 See International Renewable Energy Agency, 'Floating Foundations: A Game Changer for Offshore Wind Power', 2016. Available at www.irena.org/-/media/Files/IRENA/Agency/Publication/2016/IRENA_Offshore_Wind_Floating_Foundations_2016.pdf. See also International Energy Outlook, 'Offshore Wind Outlook 2019', World Energy Outlook Special Report, October 2019.

11 This could be in the form, for example, of the Advanced Research & Invention Agency (ARIA), a new independent research body focusing on funding high-risk, high-reward scientific research, announced by the UK government in February 2021. https://www.gov.uk/government/news/uk-to-launch-new-research-agency-to-support-high-risk-high-reward-science.

12 See International Energy Agency, 'Energy Efficiency 2018: Analysis and outlook to 2040', www.iea.org/efficiency2018/; and www.energy.gov/eere/slsc/energy-efficiency-potential-studies-catalog. A particular example of presenting the over-optimistic picture is the well-known McKinsey Curve, critiqued in chapter 5 of Helm, D., *The Carbon Crunch: How We're Getting Climate Change Wrong – And How to Fix it*, Revised and Updated. London: Yale University Press, 2015. See Enkvist, P. A., Nauclér, T. and Rosander, J., 'A cost curve for greenhouse gas reduction', February 2007, www.mckinsey.com/business-functions/sustainability our-insights/a-cost-curve-for-greenhouse-gas-reduction.

13 See, for example, Awbi, H. B., 'Ventilation for Good Indoor Air Quality and Energy Efficiency', *Energy Procedia*, 112, March 2017, pp 277–286.

BIBLIOGRAPHY

Ackerman, F., DeCanio, S. J., Howarth, R. B. and Sheeran, K., 'Limitations of integrated assessment models of climate change', *Climatic Change*, 95, 2009, pp 297–315

Awbi, H. B., 'Ventilation for Good Indoor Air Quality and Energy Efficiency', *Energy Procedia*, 112, March 2017, pp 277–286

Baltensperger, M. and Dadush, U., 'The European Union–Mercosur Free Trade Agreement: Prospects and Risks', *Policy Contribution*, 11, September 2019

Barrett, S., *Environment and Statecraft: The Strategy of Environmental Treaty-making*. Oxford: Oxford University Press, 2005

Bastin, J-F., Finegold, Y., Mollicone, D., Rezende, M., Routh, D., Zohner, C. M. and Crowther, T. W., 'The global tree restoration potential', *Science*, 365, 2019, pp 76–79

Becqué, R., Dubsky, E., Hamza-Goodacre, D. and Lewis, M., 'Europe's Carbon Loophole', September 2017

Beiser-McGrath, L. F. and Bernauer, T., 'Could revenue recycling make effective carbon taxation politically feasible?', *Science Advances,* 5(9), 18 September 2019

Berlin, I., *The Crooked Timber of Humanity: Chapters in the History of Ideas*. London: Pimlico, 2013

Berners-Lee, M., *How Bad Are Bananas? The Carbon Footprint of*

Everything, Updated and Expanded, London: Profile Books, 2020

BP, 'BP Energy Outlook: 2019 Edition', BP plc, 2019

Broome, J., *Climate Matters: Ethics in a Warming World*. New York: W. W. Norton & Company Inc., 2012

Catapult Energy Systems, 'Preparing UK Electricity Networks for Electric Vehicles', Report, 2018

Centre for Rural Economy, 'Brexit: How might UK agriculture thrive or survive?', Note No. 7, School of Natural and Environmental Sciences, Newcastle University, August 2018

Climate Change Committee, 'The Sixth Carbon Budget: The UK's Path to Net Zero', December 2020 Climate Change Committee, 'Land use: Policies for a Net Zero UK', January 2020

Climate Change Committee, 'Net Zero: The UK's contribution to stopping global warming', May 2019

Coyle, D., *GDP: A Brief but Affectionate History*. Princeton: Princeton University Press, 2014

Dasgupta, P., *Time and the Generations*. New York: Columbia University Press, 2019

Dasgupta, P. S. and Heal, G. M., *Economic Theory and Exhaustible Resources*. Cambridge: James Nisbet and Cambridge University Press, 1979

Department for Business, Energy and Industrial Strategy, 'The Energy White Paper: Powering Our Net Zero Future', December 2020, https://www.gov.uk/government/publications/energy-white-paper-powering-our-net-zero-future

Department for Business, Energy and Industrial Strategy, 'Implementing the End of Unabated Coal by 2015: Government response to unabated coal closure consultation, January 2018

Department of Energy and Climate Change, 'Implementing Electricity Market Reform (EMR): Finalised policy positions for implementation of EMR', Policy Paper, June 2014

Department for Environment, Food and Rural Affairs, 'A Green Future: Our 25 Year Plan to Improve the Environment', January 2018

Department for Transport, 'Decarbonising Transport: Setting the Challenge', March 2020

Ehrlich, P. R., *The Population Bomb: Population Control or Race to Oblivion?*. New York: Ballantine Books, 1968

European Commission, 'A new Industrial Strategy for a globally competitive, green and digital Europe', 10 March 2020

European Commission, 'Proposal for a Regulation of the European Parliament and of the Council establishing the framework for achieving climate neutrality and amending Regulation

(EU) 2018/1999 (European Climate Law)', COM(2020) 80 final, 2020/0036 (COD), 4 March 2020

European Commission, 'The European Green Deal', COM(2019) 640 final, 11 December 2019

Enkvist, P. A., Nauclér, T. and Rosander, J., 'A cost curve for greenhouse gas reduction', *McKinsey Quarterly*, February 2007

European Court of Auditors, 'The integrity and implementation of the EU ETS', Special Report, 2015

Federal Ministry for Economic Affairs and Energy, 'Commission on Growth, Structural Change and Employment: Final Report', January 2019

Fukuyama, F., *The End of History and the Last Man*. Penguin, 1989

Fullerton, D., Leicester, A. and Smith, S., 'Environmental taxes', in Institute for Fiscal Studies, *Reforming the Tax System for the 21st Century: The Mirrlees Review*. Oxford: Oxford University Press, 2011

Gallagher, K. S., Bhandary. R., Narassimhan, E. and Nguyen, Q. T., 'Banking on coal? Drivers of demand for Chinese overseas investments in coal in Bangladesh, India, Indonesia and Vietnam', *Energy, Research & Social Science*, 71, January 2021

Garnett, S., Selvidge, J., Westerberg, S. and Thompson P., 'RSPB Geltsdale – A case study of upland management', *British Wildlife*, 30(6), August 2019, pp 409–417

Goldthorpe, S. 'Potential for Very Deep Ocean Storage of CO_2

Without Ocean Acidification: A Discussion Paper', *Energy Procedia*, 114, July 2017, pp 5417–5429
Hamilton, C., *Growth Fetish*. London: Allen & Unwin, 2003
Hayhow, D. B., Eaton, M. A., Stanbury, A. J., Burns, F., Kirby, W. B., Bailey, N., Beckmann, B., Bedford, J., Boersch-Supan, P. H., Coomber, F., Dennis, E. B., Dolman, S. J., Dunn, E., Hall, J., Harrower, C., Hatfield, J. H., Hawley, J., Haysom, K., Hughes, J., Johns, D. G., Mathews, F., McQuatters-Gollop, A., Noble, D. G., Outhwaite C. L., Pearce-Higgins, J. W., Pescott, O. L., Powney, G. D. and Symes, N., 'The State of Nature 2019', The State of Nature Partnership, October 2019
Helm, D., 'Net Carbon Gain', February 2021, http://www.dieterhelm co.uk/natural-capital/environment/net-carbon-gain/
Helm, D., 'The Economic Impacts of the Coronavirus', *Environmental and Resource Economics*, (76), 2020, pp 21 38
Helm, D. 'HS2: A conclusion in search of a rationale', September 2019, http://www.dieterhelm.co.uk/regulation/transport/hs2-a-conclusion-in-search-of-a-rationale/
Helm, D., 'Power cuts and how to avoid them', August 2019, http://www.dieterhelm.co.uk/energy/energy/power-cuts-and-how-to-avoid-them/
Helm, D., 'The Systems Regulation Model', February 2019, www.dieterhelm.co.uk/regulation/regulation/the-systems-regulation-model/
Helm, D., *Green and Prosperous Land: A Blueprint for Rescuing the British Countryside*. London: HarperCollins, 2019
Helm, D., 'The Nuclear RAB Model', June 2018, http://www.dieterhelm.co.uk/energy/energy/the-nuclear-rab-model/
Helm, D., 'The Cost of Energy Review' (the Helm Review), report prepared for the Department for Business, Energy and Industrial Strategy, October 2017
Helm, D. 'Agriculture after Brexit', *Oxford Review of Economic Policy*, 33(1), 2017, pp 124–133

Helm, D., *Burn Out: The Endgame for Fossil Fuels*. London: Yale University Press, 2017

Helm, D., 'The new broadband utility and the Openreach debate', January 2016, www.dieterhelm.co.uk/regulation/communications/the-new-broadband-utility-and-the-openreach-debate/

Helm, D., *Natural Capital: Valuing the Planet*. London: Yale University Press, 2015

Helm, D., *The Carbon Crunch: How We're Getting Climate Change Wrong – And How to Fix it*, Revised and Updated. London: Yale University Press, 2015

Helm, D., 'Rethinking the economic borders of the state – Ownership, assets, and competition', *Oxford Review of Economic Policy*, 31(2), 2015, pp 168–185

Helm, D., 'European Energy Policy', in Jones, E., Menon, A. and Weatherill, S. (eds), *The Oxford Handbook of the European Union*. Oxford: Oxford University Press, 2012, chapter 39

Helm, D., 'The Sustainable Borders of the State', *Oxford Review of Economic Policy*, (27)4, 2011, pp 517–535

Helm, D., 'Regulatory Reform, Capture, and the Regulatory Burden', *Oxford Review of Economic Policy*, 22(2), Summer 2006, pp 169–185

Helm, D., *Energy, the State and the Market: British Energy Policy Since 1979*. Oxford: Oxford University Press, 2004

Helm, D., Hepburn, C. and Mash, R., 'Credible Carbon Policy', *Oxford Review of Economic Policy*, 19(3), September 2003, pp 438–450

Helm, D., Hepburn, C. and Ruta, G., 'Trade, Climate Change, and the Political Game Theory of Border Carbon Adjustments', *Oxford Review of Economic Policy*, 28(2), July 2012

Hepburn, C., Adlen, E., Beddington, J., Carter, E. A., Fuss, S., MacDowell, N., Minx, J. C., Smith, P. and Williams, C. K., 'The technological and economic prospects for CO_2 utilization and removal', *Nature*, 575, 2019, pp 87–97

Hepburn, C., 'Regulation by prices, quantities, or both: a review of instrument choice', *Oxford Review of Economic Policy*, 22(2), 2006, pp 226–247

HM Government, 'The Ten Point Plan for a Green Industrial Revolution', November 2020, https://www.gov.uk/government/publications/the-ten-point-plan-for-a-green-industrial-revolution

HM Treasury, 'Carbon Emissions Tax: Consultation', July 2020, https://assets.publishing.service.gov.uk/government/uploads/system/uploads/attachment_data/file/902737/Carbon_Emissions_Tax_-__consultation.pdf

HM Treasury, 'Net Zero Review: Interim Report', December 2020, https://www.gov.uk/government/publications/net-zero-review-interim-report

HM Treasury, 'National Infrastructure Strategy', November 2020, https://www.gov.uk/government/publications/national-infrastructure-strategy

Honohan, P., 'Should Monetary Policy Take Inequality and Climate Change into Account?', Peterson Institute for International Affairs, 19-18, October 2019

Hume, D., *Dialogues Concerning Natural Religion*, 1779

Hume, D., *An Enquiry Concerning the Principles of Morals*, 1777

International Energy Outlook, 'Offshore Wind Outlook 2019', World Energy Outlook Special Report, October 2019

International Energy Agency, 'World Energy Outlook 2018', 2018

Jackson, T., *Prosperity without Growth*, Second Edition. New York: Routledge, 2017

Jevons, W. S., *The Coal Question: An Inquiry Concerning the Progress of the Nation, and the Probable Exhaustion of Our Coal-mines*. London: Dodo Press, 2008 edition

Johnson, S. and Boswell, J., *A Journey to the Western Islands of Scotland and the Journal of a Tour to the Hebrides*, edited by P. Levi. London: Penguin Classics, 1984

Kennedy, P., *The Parliament of Man: The Past, Present and Future of the United Nations*. London: Penguin, 2006

Li, W., Li, Y., Deng, H. and Bao, L., 'Planning of Electric Public Transport System under Battery Swap Mode', *Sustainability*, 10, 2528, July 2018

Lipton, D. and Sachs, J. D., 'Creating a Market Economy in Eastern Europe: The Case of Poland', Brookings Papers on Economic Activity, 1, 1990

Lowe, S., 'Should the EU tax imported CO_2?', Centre for European Reform, September 2019

MacIntyre, A. C., 'Hume on "Is" and "Ought"', *The Philosophical Review*, 68(4), October 1959, pp 451–468

Malthus, T., *Essay on the Principle of Population*. London: Pelican Books, 1970

Meadows, D. H., Meadows, D. L., Randers, J. and Behrens, W., *The Limits to Growth: A Report for the Club of Rome's Project on the Predicament of Mankind*. New York: Universe Books, 1972

Mendel, G., *Experiments on Plant Hybridisation*. 1865

Miller, J. J., 'J. S. Mill on Plural Voting, Competence and Participation', *History of Political Thought*, 24(4), 2003, pp 647–667

Morstyn, T., Teytelboym, A. and McCulloch, M. D., 'Designing Decentralized Markets for Distribution System Flexibility', *IEEE Transactions on Power Systems*, 34(3), May 2019

Mulligan, T., 'Plural Voting for the Twenty-First Century', *The Philosophical Quarterly*, 68(271), September 2017, pp 286–306

Muûls, M., Colmer, J., Martin, R. and Wagner, U. J., 'Evaluating the EU Emissions Trading System: Take it or leave it? An assessment of the data after ten years', Grantham Institute Briefing Paper No. 21, October 2016

Natural Capital Committee, 'Advice on using nature based interventions to reach net zero greenhouse gas emissions by 2050', April 2020

Nordhaus, W. D., 'Revisiting the social cost of carbon', PNAS, 114(7), February 2017, pp 1518–1523

Nordhaus, W., *The Climate Casino: Risk, Uncertainty, and Economics for a Warming World*. New Haven: Yale University Press, 2013

Nozick, R., *Anarchy, State, and Utopia*. Oxford: Blackwell Publishing, 1974

Ofgem, 'Review of GB energy system operation', January 2021, https://www.ofgem.gov.uk/system/files/docs/2021/01/ofgem_-_review_of_gb_energy_system_operation_0.pdf

Pfeiffer, A., Millar, R., Hepburn, C. and Beinhocker, E., 'The "2 C capital stock" for electricity generation: Committed cumulative carbon emissions from the electricity generation sector and the transition to a green economy', *Applied Energy*, 179, 2016, pp 1395–1408

Pielke Jr., R. A., *The Rightful Place of Science: Disasters and Climate Change*. Phoenix: Arizona State University, 2014

Porritt, J., *Playing Safe: Science and the Environment*. London: Thames Hudson, 2000

Rural Economy and Connectivity Committee, 'Salmon farming in Scotland', Scottish Parliament Paper 432, November 2018

Smart, J. J. C. and Williams, B., *Utilitarianism: For and Against*. New York: Cambridge University Press, 1973

Smith, L. G., Kirk, G. J. D., Jones, P. J. and Williams, A. G., 'The greenhouse gas impacts of converting food production in England and Wales to organic methods', *Nature Communications*, 10, 4641, October 2019

Staedter, T., 'Big Mammals Evolved Thanks to More Oxygen', *Scientific American*, 3 October 2005, www.scientificamerican.com/article/big-mammals-evolved-thank/

Stern, N., *The Economics of Climate Change: The Stern Review*, HM Treasury. Cambridge: Cambridge University Press, January 2007

The Royal Society, 'Ammonia in a Net-zero Carbon Future: A carbon-free fuel and energy store', 2020

Tubb, C. and Seba, T., 'Rethinking Food and Agriculture 2020–2030: The Second Domestication of Plants and Animals, the Disruption of the Cow, and the Collapse of Industrial Livestock Farming', RethinkX, September 2019

Tudge, C., *In Mendel's Footnotes: An Introduction to the Science and Technologies of Genes and Genetics from the Nineteenth Century to the Twenty-Second*. London: Vintage, 2002

United Nations, 'Paris Agreement', 2015, https://unfccc.int/sites/default/files/english_paris_agreement.pdf

United Nations, 'Kyoto Protocol to the United Nations Framework Convention on Climate Change', 1998, https://unfccc.int/resource/docs/convkp/kpeng.pdf

United Nations, 'United Nations Framework Convention on Climate Change', 1992, https://unfccc.int/resource/docs/convkp/conveng.pdf

United Nations, 'Our Common Future: Report of the World Commission on Environment and Development', The Brundtland Report, 1987

Victor, D. G., *Global Warming Gridlock: Creating More Effective Strategies for Protecting the Planet*. Cambridge: Cambridge University Press, 2011

Victor, D. G., *The Collapse of the Kyoto Protocol and the Struggle to Slow Global Warming*. Princeton: Princeton University Press, 2004

Wang, M., Han, J., Dunn, J. B., Cai, H. and Elgowainy, A., 'Well-to-wheels energy use and greenhouse gas emissions of ethanol from corn, sugarcane and cellulosic biomass for US use', *Environmental Research Letters*, 7, 2012, pp 1–13

Widger, P. and Haddad, A. M., 'Evaluation of SF6 Leakage from Gas Insulated Equipment on Electricity Networks in Great Britain', *Energies*, 11, 2037, August 2018

Yergin, D., *The Prize: The Epic Quest for Oil, Money and Power*. New York: Free Press, 1991

INDEX